NATIONAL SCIENCE AND TECHNOLOGY COUNCIL

NATIONAL TRANSPORTATION STRATEGIC RESEARCH PLAN

Committee on Technology
Subcommittee on Transportation Research and Development

Prepared by
United States Department of Transportation
Research and Special Programs Administration
John A. Volpe National Transportation Systems Center
Transportation Strategic Planning and Analysis Office
55 Broadway
Cambridge, MA 02142

May 2000

REPRODUCED BY: **NTIS.**
U.S. Department of Commerce
National Technical Information Service
Springfield, Virginia 22161

The purpose of this report is to help Congress and the Administration establish national transportation research and technology priorities and coordinated research activities. The report is intended to provide a planning framework for Federal, State and local governments; academia; and industry in supporting national transportation goals. It also conveys to the science and technology community the types of research and research priorities being sponsored and considered by the Federal agencies. The Administration is committed to a broad range of high-priority investments (including science and technology); to deficit reduction; and to a smaller, more efficient Federal Government. These commitments have created a very challenging budget environment—requiring difficult decisions and a well thought out strategy to ensure the best return for the Nation's taxpayers. As part of this strategy, this document does not represent the final determinant in an overall Administration budget decision-making process. Research programs presented in this report will have to compete for resources against many other high-priority Federal programs. If these programs compete successfully, they will be reflected in future Administration budgets.

PROTECTED UNDER INTERNATIONAL COPYRIGHT
ALL RIGHTS RESERVED
NATIONAL TECHNICAL INFORMATION SERVICE
U.S. DEPARTMENT OF COMMERCE

Technical Report Documentation Page

1. Report No. DOT-T-00-04	2. Government Accession No.	3. Recipient's Catalog No.
4. Title and Subtitle NATIONAL TRANSPORTATION STRATEGIC RESEARCH PLAN		5. Report Date May 2000
		6. Performing Organizational Code
7. Author(s)		8. Performing Organization Report No.
9. Performing Organization Name and Address United States Department of Transportation Research and Special Programs Administration John A. Volpe National Transportation Systems Center Transportation Strategic Planning and Analysis Office 55 Broadway, Cambridge, MA 02142		10. Work Unit No. (TRAIS)
		11. Contract or Grant No.
12. Sponsoring Agency Name and Address U.S. Department of Transportation's Technology Sharing Program 400 Seventh Street, S.W., Room 8417 Washington, DC 20590		13. Type of Report and Period Covered May 2000
		14. Sponsoring Agency Code DOT/RSPA

15. Supplementary Notes
Released in cooperation with the U.S. Department of Transportation's Technology Sharing Program

16. Abstract
This is one of two implementation documents for the National Transportation Science and Technology Strategy. It deals with the areas of basic or enabling research which are necessary to provide the technological foundations for new and improved transportation systems. The Plan considers the roles of a broad range of parties involved in the transportation enterprise: Federal government agencies; State, local, and tribal agencies; academic institutions, and the private sector. It focuses on seven areas of enabling research: human performance and behavior; advanced materials and structures; computer, information, and communication systems; energy, propulsion, and environmental engineering; sensing and measurement; analysis, modeling, design and construction tools; and social and economic policy issues.

17. Key Words Enabling Research; Human Performance; Advanced Materials; Sensors; Computer Analysis; Transportation Design; Economic Policy	18. Distribution Statement **Availability is unlimited. Document is being released for sale to the U.S. public through the National Technical Information Service, Springfield, Virginia 22161**		
19. Security Classif. (of this report) **Unclassified**	20. Security Classif. (of this page) **Unclassified**	21. No. of Pages 63	22. Price

Form DOT F 1700.7 (8-72)   Per FORM PRO/Delrina 04/04/94   Reproduction of completed page authorized

THE DEPUTY SECRETARY OF TRANSPORTATION
WASHINGTON, D.C. 20590

Dear Colleague:

We are pleased to present the National Science and Technology Council's *National Transportation Strategic Research Plan*. This plan, an update of the 1998 *Transportation Strategic Research Plan*, is a cooperative product of the Federal agencies that participate in transportation-related research, strengthened by external review by the transportation community. It provides a broad overview of the transportation research agenda across the entire Federal government and highlights the activities of specific agencies and achievements in particular areas. It is this research and technology development that will lay the foundation for the technical breakthroughs that will be required in the 21st century, if we are to meet the nation's transportation needs in a safe and environmentally benign way. The plan also provides a framework within which to set directions for future initiatives and cooperation.

As the United States enters the 21st century, the continuing commitment to high-quality enabling research will make the transportation system safer, more secure, more efficient, and more reliable. It also will generate other benefits, including:

- Development of a more highly skilled, more efficient, and better trained public and private sector transportation work force;
- Support for advances in other areas of concern, such as the environment;
- Production of a stronger, more capable national defense network; and
- Greater partnership efforts and coordination of transportation issues with the private, non-profit, and academic sectors.

The ongoing leadership that our nation demonstrates in transportation-related research includes international needs as well as domestic concerns.

President Clinton has challenged us to "... harness the remarkable forces of science and technology that are remaking our world...." The following pages rise to that challenge by providing a foundation of enabling research for the innovations that will shape transportation in the 21st century.

Sincerely,

Mr. Mortimer Downey
Chair

Dr. Henry Kelly
White House Co-Chair

Committee on Technology, Subcommittee on Transportation R&D
National Science and Technology Council

NSTC COMMITTEE ON TECHNOLOGY
SUBCOMMITTEE ON TRANSPORTATION R&D

Chair: Mr. Mortimer L. Downey	Deputy Secretary, U.S. Department of Transportation
Vice-Chair: Mr. Samuel Venneri	Associate Administrator for Aero-Space Technology, National Aeronautics and Space Administration
White House Co-Chair: Mr. Henry Kelly	Assistant Director for Technology, Office of Science and Technology Policy, Executive Office of the President
Executive Director: Dr. E. Fenton Carey	Associate Administrator for Innovation, Research and Education, Research and Special Programs Administration, U.S. Department of Transportation

Members:

Ms. Cheryl Shavers	Under Secretary for Technology, U.S. Department of Commerce
Mr. Peter "Jack" Basso	Assistant Secretary for Budget and Programs, U.S. Department of Transportation
Dr. Charles Holland	Director, Information Systems, Defense Research and Engineering, U.S. Department of Defense
Mr. Eugene A. Conti, Jr.	Assistant Secretary for Transportation Policy, U.S. Department of Transportation
Mr. Michael Deich	Associate Director for General Government and Finance, Office of Management and Budget
Mr. Richard T. Farrell	Assistant Administrator for Policy, Environmental Protection Agency
Mr. Dan W. Reicher	Assistant Secretary for Energy Efficiency and Renewable Energy, U.S. Department of Energy
Dr. Pricilla P. Nelson	Director, Civil and Mechanical Systems Division, National Science Foundation
Consultant: Dr. Richard R. John	Director, Volpe National Transportation Systems Center, U.S. Department of Transportation

Ad Hoc Members:

Ms. Julie A. Cirillo	Acting Deputy Administrator, Federal Motor Carrier Safety Administration
Ms. Kelley S. Coyner	Administrator, Research and Special Programs Administration
Ms. Nuria I. Fernandez	Acting Administrator, Federal Transit Administration
Ms. Jane F. Garvey	Administrator, Federal Aviation Administration
Mr. Clyde J. Hart, Jr.	Administrator, Maritime Administration
Adm. James M. Loy	Commandant, United States Coast Guard
Ms. Rosalyn G. Millman	Acting Administrator, National Highway Traffic Safety Administration
Ms. Jolene M. Molitoris	Administrator, Federal Railroad Administration
Dr. Ashish Sen	Director, Bureau of Transportation Statistics
Mr. Kenneth R. Wykle	Administrator, Federal Highway Administration

TABLE OF CONTENTS

EXECUTIVE SUMMARY iii

1. **INTRODUCTION** 1
   BACKGROUND 1
   PURPOSE 2
   METHODOLOGY AND SCOPE 2

2. **FEDERAL TRANSPORTATION-RELATED RESEARCH AND DEVELOPMENT** 5
   FEDERAL ENABLING RESEARCH 5
   CATEGORIES OF ENABLING RESEARCH 5
   THE FEDERAL RESEARCH COMMUNITY 6
   R&D FUNDING OVERVIEW 9

3. **"BREAKTHOUGH" RESEARCH DIRECTIONS** 13
   BREAKTHROUGH RESEARCH 13
   NANOTECHNOLOGY 13
   BIOFUELS 15
   COMPLEX SYSTEMS AND HIGH-CONFIDENCE SOFTWARE 17

4. **AREAS OF ENABLING RESEARCH** 19
   OVERVIEW 19
   HUMAN PERFORMANCE AND BEHAVIOR 22
   ADVANCED MATERIALS AND STRUCTURES 25
   COMPUTER, INFORMATION, AND COMMUNICATIONS SYSTEMS 28
   ENERGY, PROPULSION, AND ENVIRONMENTAL ENGINEERING 36
   SENSING AND MEASUREMENT 40
   ANALYSIS, MODELING, DESIGN, AND CONSTRUCTION TOOLS 42
   SOCIAL AND ECONOMIC POLICY ISSUES 45

5. **FUTURE DIRECTION AND PRIORITIES** 47
   FOCUS AREAS FOR HUMAN PERFORMANCE AND BEHAVIOR 48
   FOCUS AREAS FOR ADVANCED MATERIALS AND STRUCTURES 49
   FOCUS AREAS FOR COMPUTER, INFORMATION, AND COMMUNICATION SYSTEMS. ... 50
   FOCUS AREAS FOR ENERGY, PROPULSION, AND ENVIRONMENTAL ENGINEERING 51
   FOCUS AREAS FOR SENSING AND MEASUREMENT 52
   FOCUS AREAS FOR ANALYSIS, MODELING, DESIGN, AND CONSTRUCTION TOOLS 53

APPENDIX A: PARTNERSHIP INITIATIVES 55

APPENDIX B: LIST OF ACRONYMS 61

SELECTED BIBLIOGRAPHY 63

LIST OF FIGURES

FIGURES		PAGE
1.	Percentage of FY1998 Transportation-Related Enabling Research Performed by Each Relevant Agency	iv
2.	FY1998 Transportation-Related R&D Budget Authority Categorized by Area Of Enabling Research	iv
1-1.	Elements of the Transportation Science and Technology Strategy	2
2-1.	FY1998 Total Federal R&D Budget Authorization for Selected Agencies	9
2-2.	Allocation of FY1998 R&D Budget Authorization Among Selected Agencies	10
2-3.	FY1998 Transportation-Related Federal R&D Budget Authorization for Selected Agencies	10
2-4.	FY1998 Transportation-Related R&D Budget Authorization for Selected Agencies	11
2-5.	Percentage of FY1998 Agency R&D Budget Authorization Identified with Transportation-Related Enabling Research	11
4-1.	FY1998 Total Transportation-Related R&D Budget Authorization by Category	19
4-2.	FY1998 Transportation-Related R&D Budget Authorizations Categorized by Area of Enabling Research (Dept. of Defense only)	20
4-3.	FY1998 Transportation-Related R&D Budget Authorization for Selected Agencies other than Dept. of Defense Categorized by Area of Enabling Research	20
4-4.	FY1998 Transportation-Related R&D Budget Authorization Categorized by Area of Enabling Research	21
4-5.	FY1998 Transportation-Related R&D Budget Authority by Agency and Area of Enabling Research for Agencies other than DoD	21

EXECUTIVE SUMMARY

The NSTC's *National Transportation Science and Technology Strategy,* issued in April 1999, has four key elements: Strategic Planning and Assessment, Partnership Initiatives, Enabling Research, and Education and Training. This Transportation Strategic Research Plan addresses the Enabling Research element. It incorporates R&D activities with clear potential relevance to one or more transportation modes or functions, regardless of the objectives for which it is conducted or the performing agency. Enabling research includes activities described under three Federal budget categories. For civil agencies, these are Basic Research, Applied Research, and Development. The comparable Defense Department terms are 6.1 (Basic Research), 6.2 (Applied Research), and 6.3 (Advanced Technology Demonstration).

All of the agencies participating in the NSTC Transportation R&D Subcommittee conduct enabling research that has clear direct or long-term application to the nation's transportation enterprise. This document presents a broad overview and categorization of that research, including identification of many of the program areas being addressed. The intent is to ensure that the transportation enterprise takes full advantage of this research, and that the affected Federal agencies coordinate their efforts and maximize synergies among their respective efforts.

This document, the *National Transportation Strategic Research Plan,* builds on the initial *Strategic Research Plan* issued in 1999 to present updated budget authorization data and other information on Federal enabling research. It differs from the 1999 plan in three ways:

It incorporates more recent data

It includes a seventh category of enabling research, Social and Economic Policy Issues

It contains a section on critical "breakthrough" research areas

This plan draws on an extensive examination of government documents, publications, and Internet Web sites. The principal data resource used is the RaDiUS (Research and Development in the United States) database established by the Office of Science and Technology Policy. It contains R&D budget authority data and project information covering 25 Federal agencies and virtually all of the more than $70 billion annual Federal research allocation.*

---

* RaDiUS has been developed by RAND, in cooperation with the National Science Foundation (NSF), to support the work of RAND's Science & Technology Policy Institute, the federally funded research and development center serving the White House Office of Science and Technology Policy and the National Science and Technology Council. The data in RaDiUS come from many sources located throughout the Federal Government. Among these sources are the Catalog of Federal Domestic Assistance (CFDA); USDA's Current Research Information System (CRIS); HHS's Computer Retrieval of Information on Scientific Projects (CRISP) and Information for Management, Planning, Analysis, and Coordination (IMPAC) system; DoD's R-1 and R-2 Budget Exhibits and Technical Effort And Management System (TEAMS) [formerly the Work Unit Information Systen (WUIS)]; DOE's laboratory information system; the Federal Assistance Awards Data System (FAADS); the Federal Procurement Data System (FPDS); OMB's MAX system; DVA's R&D Information System (RDIS); NSF's Science and Technology System (STIS); and NASA's 507 System.

Analysis of this database indicates that in fiscal year 1998 transportation-related enabling research comprised approximately $5.1 billion annually, equivalent to 9 percent of the cumulative $56 billion in R&D budget authority represented by the agencies that are represented on the Transportation R&D Subcommittee.

The fraction of agency R&D identified as (1) directly or potentially relevant to transportation and (2) enabling in nature (rather than mission-driven or developmental) ranged from 5 percent (for the Department of Commerce) to more than 60 percent (for the Department of Transportation). However, since agencies such as the Department of Defense, NASA and the National Science Foundation have relatively large R&D budgets, their impact in providing enabling research is large even where it comprises a small part of their total program. Figure 1 indicates the portion of enabling research provided by each of the selected agencies; Defense represents 38 percent of the total, followed by Department of Energy (25 percent) and NASA (11 percent).

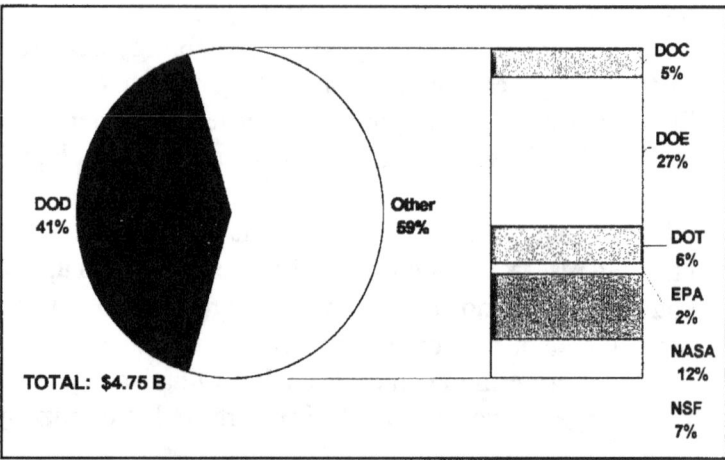

Figure 1. Percentage of FY1998 Transportation-Related Enabling Research Performed by Each Relevant Agency.

Budgets for enabling research were subdivided in terms of the seven categories defined in the *National Transportation Science and Technology Strategy*. These are:
- Human Performance and Behavior
- Advanced Materials and Structures
- Computer, Information, and Communication Systems
- Energy, Propulsion, and Environmental Engineering
- Sensing and Measurement
- Analysis, Modeling, Design, and Construction Tools
- Social and Economic Policy Issues

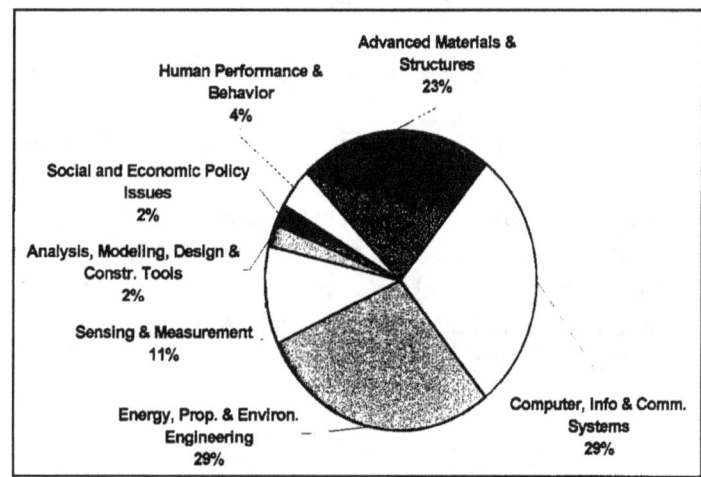

Figure 2. FY1998 Transportation-Related R&D Budget Authority Categorized by Area of Enabling Research.

Figure 2 shows the percentage of total enabling research allocated to each of these categories.

This plan is the first step in an ongoing, evolving process to guide and stimulate strategic research in transportation. The emphasis in this initial plan has been development of a basic scan of ongoing Federal research that, regardless of the purpose for which it was undertaken or the performing agency, can potentially contribute to innovation in transportation applications. As such, this plan presents a context and framework for increasing the synergy among Federal agency research programs, and heightening awareness and utilization of the results of that research by states, localities, and the private sector.

1. INTRODUCTION

BACKGROUND

The high level of mobility and affordable access made possible by the U.S. transportation system is critical to America's economic well-being and quality of life. The nation's sheer physical size makes a high-quality, high performance, reliable and efficient national transportation system central to our domestic and global competitiveness. However, that system faces severe challenges. A growing and changing population demands higher capacity, greater efficiency and ever-improving levels of service. At the same time, adverse environmental impacts as well as death and injury rates have been significantly reduced in recent decades, but are still at levels unacceptable to most people.

The FY 2000 budget of the United States notes that "Science and technology are principal agents of change and progress, with over half of the Nation's economic productivity growth in the last 50 years attributable to technological innovation and the science that supports it." This statement has particular applicability to the transportation sector, which has been dramatically transformed by technology for more than a century and a half. The steam engine was a central factor in the industrial revolution and global expansion. Railroads, drawing on enabling technologies such as the steam engine, telegraph, and innovative civil engineering, remade America in the mid-1800s. The electric streetcar made suburbs possible in the late 19th century. The early years of the 20th century saw the development and vigorous exploitation of the internal combustion engine. The "superhighway" now makes possible a level of personal mobility hardly imagined a century ago. More recently, the jet engine, the "mega-ship," and containerized freight have made their appearance. Modern transportation is virtually defined by these technologies.

The needs of the future can only be met by continuing advances in a wide range of technologies and their timely incorporation into transportation applications. Research and development is an essential component in the innovation process, and the Federal government is a major participant in basic and applied research as well as subsequent development. A significant portion of research and development (R&D) has the potential to enable significant technological innovations in transportation, though much of it is performed to meet agency responsibilities in other spheres. It is especially important that the full spectrum of government R&D be identified, regardless of the agency involved or the purpose of the research, to assure that the transportation system will realize maximum benefits.

Accordingly, the National Science and Technology Council (NSTC) Subcommittee on Transportation R&D has undertaken an assessment of the full range of Federal research potentially relevant to transportation. This topic–Enabling Research– represents one of the four basic elements of the NSTC *Transportation Science and Technology Strategy*, as suggested in Figure 1-1. Together with the other three elements, the result is a comprehensive approach to structuring and coordinating Federal transportation science and technology activities.

The Strategic Planning and Assessment element provides the outcome goals, overall coordination, and assessment that are the framework for the other parts of the strategy. Strategic Partnership Initiatives seek to combine and leverage resources from multiple agencies and the private sector to expedite near-term application of emerging technological advances. The fourth element, Education and Training, ensures a continuing investment in the human capital of those who plan, design, construct, operate and maintain the transportation system.

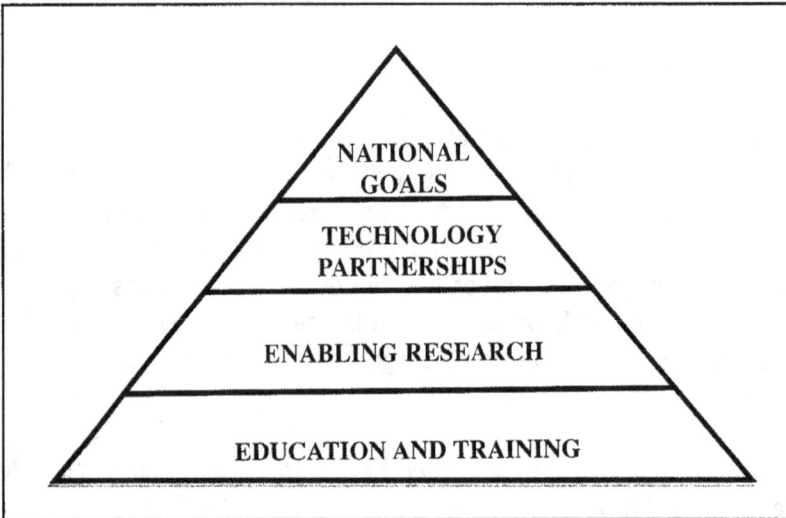

Figure 1-1. Elements of the Transportation Science and Technology Strategy.

PURPOSE

This plan is the product of an overview assessment of transportation-related enabling research now being conducted by Federal agencies. Its purpose is to describe, from a high-level perspective, the broad range of Federal research now underway that can enable the advances needed in the next century, and to provide the foundation necessary for coordination and full exploitation of all transportation-related R&D.

METHODOLOGY AND SCOPE

This plan is based on an extensive literature search of government documents, publications, and Internet Web sites. This includes a review of materials provided by the agencies and institutions that are part of the Federal research community. However, in order to achieve both completeness and consistency across agencies, the budget data reported here are all derived from the RaDiUS (Research and Development in the United States) database. This resource, established by the Office of Science and Technology Policy and implemented through the RAND Corporation, contains R&D budget authority and project information covering 25 Federal agencies. It involves more than 1,500 programs and nearly 300,000 separate grants and contracts. It tracks virtually all of the more than $70 billion annual Federal research expenditure.[1,2] (Wherever the term "funding" is used in this document, it refers to budget authority.)

---

[1] More detailed analysis of specific aspects of this topic is facilitated by the numerous Internet World Wide Web sites for the relevant agencies and other organizations that track research activities.

[2] The data in RaDiUS come from many sources located throughout the Federal Government. Among these sources are the Department of Defense R-1 and R-2 Budget Exhibits and Technical Effort And Management System; the Department of Energy laboratory information system; the Federal Procurement Data System; the Office of

Based on the literature search, the scope of this plan includes research conducted by the Departments of Transportation (DOT), Energy (DOE), Commerce (DOC) and Defense (DoD), as well as National Aeronautics and Space Administration (NASA), the Environmental Protection Agency (EPA), and National Science Foundation (NSF). These agencies represent a major portion of all Federal R&D, with the remainder conducted largely by the National Institutes of Health. The Federal budget process established by the Office of Management and Budget (OMB), which is embodied in RaDiUS, characterizes R&D as basic research, applied research and development. For this initial examination, all three categories were included as potentially representing enabling research, though development of a technology for a non-transportation application may still not be close to market-readiness for application in transportation.

Based on project or program titles in the RaDiUS budget data—supplemented where possible by other resources—the research and development judged to be transportation-related was identified. Then it was grouped quantitatively for each agency in terms of the seven enabling research categories defined in the *Transportation Science and Technology Strategy* (See Section 2.1 and Section 3). Illustrative examples of specific R&D projects for the various agencies were also presented for each category, based on information from a variety of resources including organization R&D plans and World Wide Web sites.

The process is necessarily imprecise. Given the great extent and breadth of Federal R&D, and the overview nature of this plan, the assessment did not include examination of detailed program and project text descriptions. In some instances, the titles of research projects as reported in the RaDiUS database do not provide a clear indication of the projects' transportation-related components. In other instances, the research categories focus on the application rather than the technical area or discipline, and do not have a direct correlation with the seven categories.

It is not always possible to make a clear distinction between enabling research and elements of the strategic partnership initiatives set forth in the NSTC *Strategy*. The initiatives, having a more near-term focus, can draw on funds appropriated for R&D, advanced development, evaluation and test, and implementation. Accordingly, there is some overlap of the enabling research addressed in this plan and the research portions of the initiatives. Appendix A provides a list and brief description of the initiatives.

Another source of uncertainty is that different types of research may be conducted within one overall program or project. For example, research on human performance and behavior is often subsumed in overall system development projects, and therefore is not visible in the RaDiUS tables. The overall research titles and RaDiUS categories do not indicate the human performance and behavior component to the research. Thus, the judgments as to relevance to transportation, and, to a lesser degree, the appropriate category of enabling research, are inherently somewhat uncertain. However, for the purpose of providing a broad picture of transportation-related research, indicating the major participating agencies, and assessing the distribution among R&D categories, the data presented in this plan should be fully satisfactory.

Management and Budget MAX system; the National Science Foundation Science and Technology System; and NASA's 507 System.

A central resource in shaping this plan has been a workshop conducted for the NSTC Technology Committee, with DOT funding by the National Research Council (NRC)/Transportation Research Board (TRB) in September 1998. This event brought together knowledgeable individuals from academia, the private sector, and Federal agencies to identify (1) important research already being funded within the government–not necessarily linked to transportation–which could be leveraged by the transportation sector; (2) areas that could lead to breakthroughs in transportation technologies, concepts and systems; and (3) research areas that warrant funding priority. This external review enables this plan to be considered national in scope, rather than reflecting merely the perspective of the federal government.

2. FEDERAL TRANSPORTATION-RELATED RESEARCH AND DEVELOPMENT

FEDERAL ENABLING RESEARCH

In this report, the term "Enabling Research" is defined as research and development activities with clear potential relevance to one or more transportation modes or functions, regardless of the objectives for which it is conducted or the performing agency. Enabling research includes activities described under three Federal budget categories. For civil agencies, these are Basic Research, Applied Research, and Development. The comparable Defense Department terms are 6.1 (Basic Research), 6.2 (Applied Research), and 6.3 (Advanced Technology Demonstration). Since the focus of this document is on research relevant to transportation, but for which no specific application has been identified, enabling research is not expected to yield operational implementation for at least five years.

As set forth in the NSTC *Transportation Science and Technology Strategy* (September 1997), the following characteristics make enabling research appropriate for Federal involvement and funding:

- Supports long-term national transportation goals;
- Has benefits that are too diverse for a single company to recover and profit from its investment;
- Is associated with cost or risk that is beyond the capacity of any individual company; and
- Generates benefits that will begin to be realized too far in the future to pass the threshold of private investment criteria.

CATEGORIES OF ENABLING RESEARCH

The NSTC *National Transportation Science and Technology Strategy* (April 1999) defines seven categories of enabling research:

- Human Performance and Behavior;
- Advanced Materials and Structures;
- Computer, Information, and Communication Systems;
- Energy, Propulsion, and Environmental Engineering;
- Sensing and Measurement;
- Analysis, Modeling, Design, and Construction Tools; and
- Social and Economic Policy Issues

These categories are described and categorized in Section 3.

THE FEDERAL RESEARCH COMMUNITY

The transportation research community includes participants from the public, private, and academic sectors. These organizations work individually and in partnership to perform fundamental and advanced research with transportation implications.

The US Department of Transportation has the most direct and explicit Federal responsibility for transportation. Nearly all DOT Operating Administrations carry out important transportation research directed toward improved performance of their agency missions and, more generally, to strengthen the national transportation system. However, several other agencies also conduct a substantial amount of transportation-related research as part of their respective missions. The broad thrust of these activities is described below.

Department of Commerce. Much of the transportation-related R&D conducted under DOC is performed by the National Institute for Standards and Technology (NIST), which has as its primary mission, the promotion of US economic growth by working with industry to develop and apply technology, measurements, and standards. This research is particularly visible in its lead role in the Partnership for a New Generation of Vehicles (PNGV) program and in its construction materials research, much of which is relevant to transportation. Research conducted by the National Weather Service is also important to the entire transportation enterprise.

Department of Defense. The Department of Defense accounts for nearly half of all Federal R&D. As a large contributor to R&D for the Federal government, and with a mission inherently requiring mobility and logistics, many transportation-related advancements are achieved through their R&D efforts. These include DoD participation in the Department of Commerce-led PNGV research and other surface vehicle technologies, aviation/aeronautical technology, ship design and propulsion, satellite positioning and communications, design tools, and information technologies.

Many defense programs include significant consideration of human performance in operation of aircraft. The Defense Advanced Research Project Agency has provided substantial funding in areas including electric vehicles and maritime technologies. The Technology Reinvention Project grants awards directed toward commercialization of innovative transportation-related applications.

Department of Energy. The Department of Energy is primarily concerned with energy conservation and reduction of petroleum dependence, so it is naturally involved with transportation. The transportation energy program R&D budget request for FY 2000 is $306 million; more than half of this is associated with PNGV. It is responsible for the major part of Federal PNGV funding. Overall, the transportation energy program emphasizes alternative fuels and electric propulsion. DOE has 20 major laboratories and similar facilities. Many have strong capabilities in materials, energy conversion and storage, electronics, instrumentation, and system and data analysis.

Department of Transportation. Most of the DOT operating administrations have R&D programs in support of their various missions. Due to their explicit transportation focus and the variety of the topics pursued, each is described separately below.

Bureau of Transportation Statistics (BTS). BTS is not formally included within the DOT Research and Technology Budget Submission. However, its funding is authorized in the Research section of the Transportation Equity Act for the 21st Century. BTS provides the critical knowledge and understanding of our transportation system that is needed in the assessment of research needs and opportunities, as well as in formulation of policy.

Federal Aviation Administration (FAA). FAA's research includes security technology, programs addressing weather, aircraft safety, and the role of human performance and behavior in safe performance flight crew, maintenance and controller functions. Another major R&D component focuses on operation, maintenance and renewal of the air traffic control system. Airport research involves advanced payment design methodologies and testing as well as important safety programs (e.g., crash and fire rescue, runway and taxiway markings and lighting).

Another strategic research area involves partnerships with NASA, DoD, states and industry. This assures low-cost access to space through improved technology and operations for the rapidly growing commercial space transportation sector. It also relates to the safe integration of new spaceports and routine launch operations of reusable vehicles into the National Air Space Management System.

Federal Highway Administration (FHWA). FHWA has a program to assure the development and widespread application of advanced technology and innovative approaches in ongoing operation, maintenance and renewal of the nation's highway systems, as well as safety and environmental enhancement.

Federal Motor Carrier Safety Administration (FMCSA). Formerly an office within FHWA, FMCSA became a separate DOT operating administration on January 1, 2000. It is responsible for the issuance, administration, and enforcement of Federal laws and regulations addressing motor carrier and driver safety, hazardous materials, and drug and alcohol testing requirements.

Intelligent Transportation Systems Joint Program Office (ITS JPO). The DOT Intelligent Transportation System Program, managed by the ITS Joint Program Office and housed in FHWA, is fostering and supporting application of advanced information technologies to improve surface transportation mobility, capacity, safety and environmental compatibility. Major program elements include development of an intelligent vehicle and supporting deployment of information infrastructure for rural and urban highway applications, commercial vehicle operations, and public transit systems.

Federal Railroad Administration (FRA). FRA conducts a railroad safety R&D program that addresses human factors, rolling stock and components, track and structures, track-train interaction, train control, highway-railroad grade crossings, hazardous materials, train occupant protection, railroad system safety, and R&D facilities and equipment. FRA also sponsors technology development and demonstration projects to facilitate introduction of high-speed passenger rail services.

Federal Transit Administration (FTA). FTA research is aimed at stimulating application of technological innovation in transit system operations, including programs such as development and testing of hybrid-electric buses and fuel cell and battery-powered propulsion systems.

Maritime Administration (MARAD). MARAD does not currently have an explicit research budget, but it does participate actively in several important cooperative programs to advance innovation in shipbuilding and marine operations.

National Highway Traffic Safety Administration (NHTSA). NHTSA R&D conducts research to improve vehicle crash avoidance and crashworthiness; collects and analyzes crash data regarding the vehicle crashes through the National Center for Statistics and Analysis; in conjunction with FHWA, conducts the intelligent vehicle initiative research; develops national guidelines for crash injury mechanisms through the National Transportation Biomechanics Research Center; and with FHWA, has developed and is building the National Advanced Driving Simulator.

Research and Special Programs Administration (RSPA). The RSPA agenda emphasizes transportation strategic planning and system assessment in support of DOT and NSTC. In addition, RSPA conducts continuing R&D in support of its responsibilities in pipeline safety and transport of hazardous materials.

US Coast Guard (USCG). Coast Guard research is focused on technologies, materials and human factors topics directly related to improvement of mission performance. USCG is partnering with the Navy, DOC, and other DOT modes to design, develop and test a standard fuel cell propulsion system for marine and other heavy-duty vehicular applications.

Intermodal Cooperation on Cross-Cutting Research. The Department conducts a variety of activities, typically coordinated by the Office of the Secretary or the Research and Technology Coordinating Council, that address cross-modal R&D and related issues such as human factors and climate change.

Environmental Protection Agency. EPA research bearing on transportation primarily involves abatement, control and compliance, and specific programs on air and water quality.

National Aeronautics and Space Administration. NASA has a long and distinguished history of aeronautical R&D. While topics such as propulsion, aerodynamics and control systems have predominated, NASA is also now emphasizing aviation safety and air traffic management technology. NASA is also exploring a new generation of environmentally compatible and economically feasibly subsonic and high-speed civil transport aircraft.

National Science Foundation. The National Science Foundation is an independent Federal agency responsible for promoting science and engineering through programs that invest over $3.3 billion per year in almost 20,000 research and education projects in science and engineering. Its mission includes initiation and support of scientific and engineering research and programs to strengthen scientific and engineering research potential.

R&D FUNDING OVERVIEW

The seven Departments and agencies described above cumulatively, in fiscal year 1998, had an R&D budget authority of approximately $57.5 billion, a figure that has been relatively constant during the 1990s. This represents 80 percent of total Federal R&D for all organizations. The distribution of all R&D funding among these agencies is presented in Figures 2-1 and 2-2. The dominance of the Department of Defense, which receives two-thirds of the total FY1998 R&D budget authority for the selected agencies, is clear. R&D identified as transportation-related enabling research (as defined in this plan) totals $4.75 billion. In addition, a further $3.13 billion is invested in R&D of transportation relevance in areas other than the seven defined in the *National Transportation Science and Technology Strategy* as enabling research.[3] The data for the subset of are shown in Figures 2-3, 2-4 and 2-5. Although the patterns are similar, there are significant differences in agency totals due to the portion of each agency's R&D funding relevant to transportation. This can be seen in Figure 2-5, which shows the percentage of each agency's R&D budget identified as transportation enabling research.[4]

Note: The material in the following figures is derived from the RaDiUS database for FY 1998 budget authority. RaDiUS provides the most comprehensive resource available for identifying multi-agency Federal research expenditures at present, but full consistency across agencies is not possible due to variability among agencies in terms of definitions, reporting categories, and in some cases, methods of calculation.

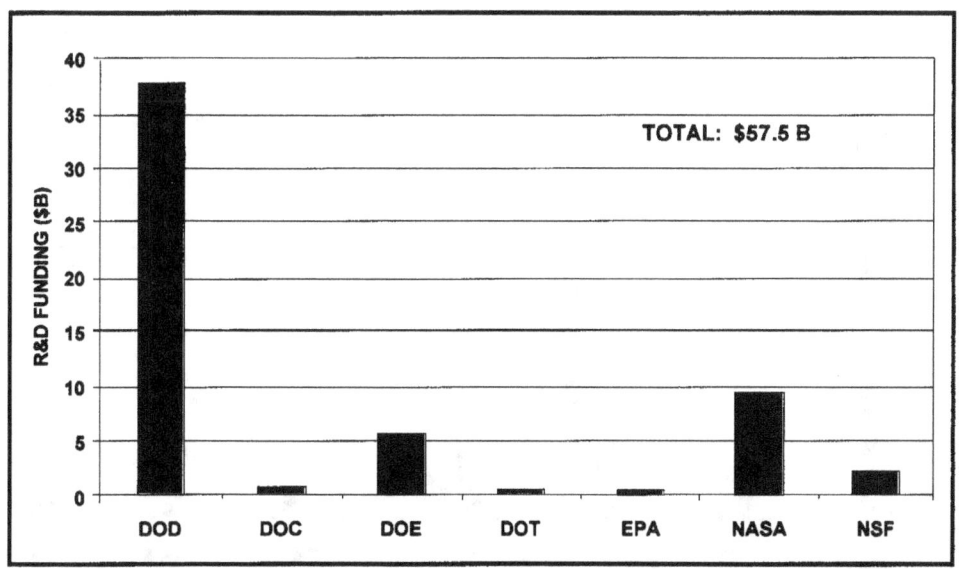

Figure 2-1. FY1998 Total Federal R&D Budget Authorization for Selected Agencies.

---

[3] This consists of $.63 B for electrical and electronic technologies, $1.31 B for vehicle-related technologies and systems, $.97 B for Knowledge Base, R&D Infrastructure, and Technology Transfer, and $.22 B for weather systems.

[4] The value of 61 percent for DOT in Figure 2-6 reflects direct or near-term mission-related R&D that, while clearly transportation-related, fall outside the definition of enabling research.

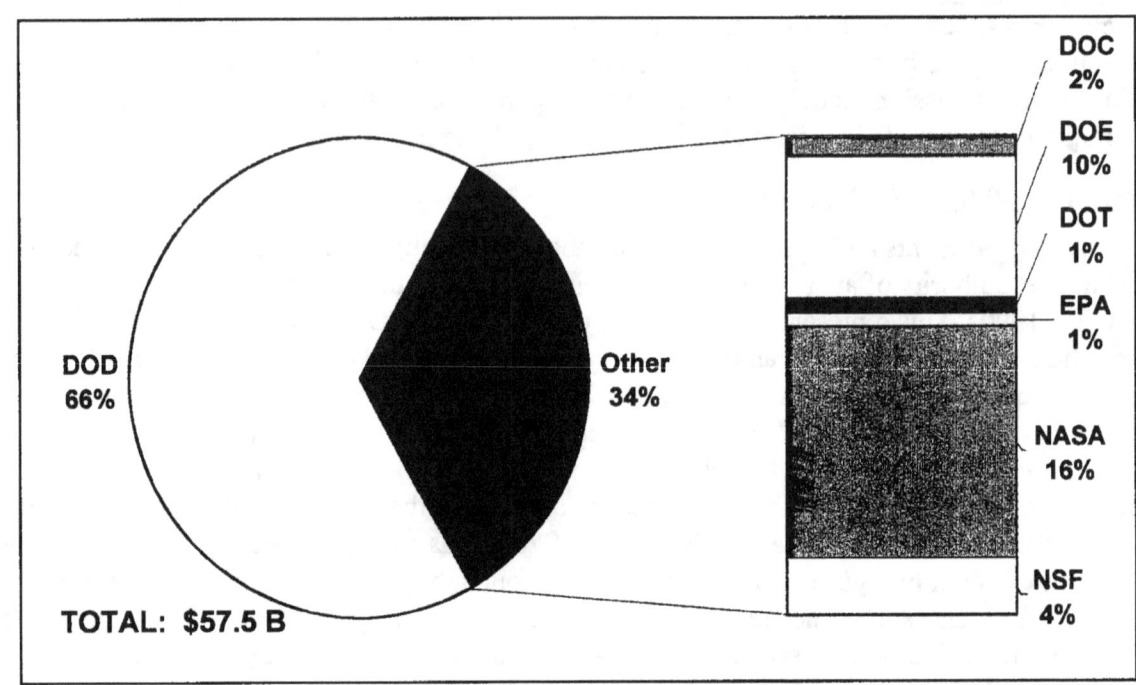

Figure 2-2. Allocation of FY1998 R&D Budget Authorization Among Selected Agencies.

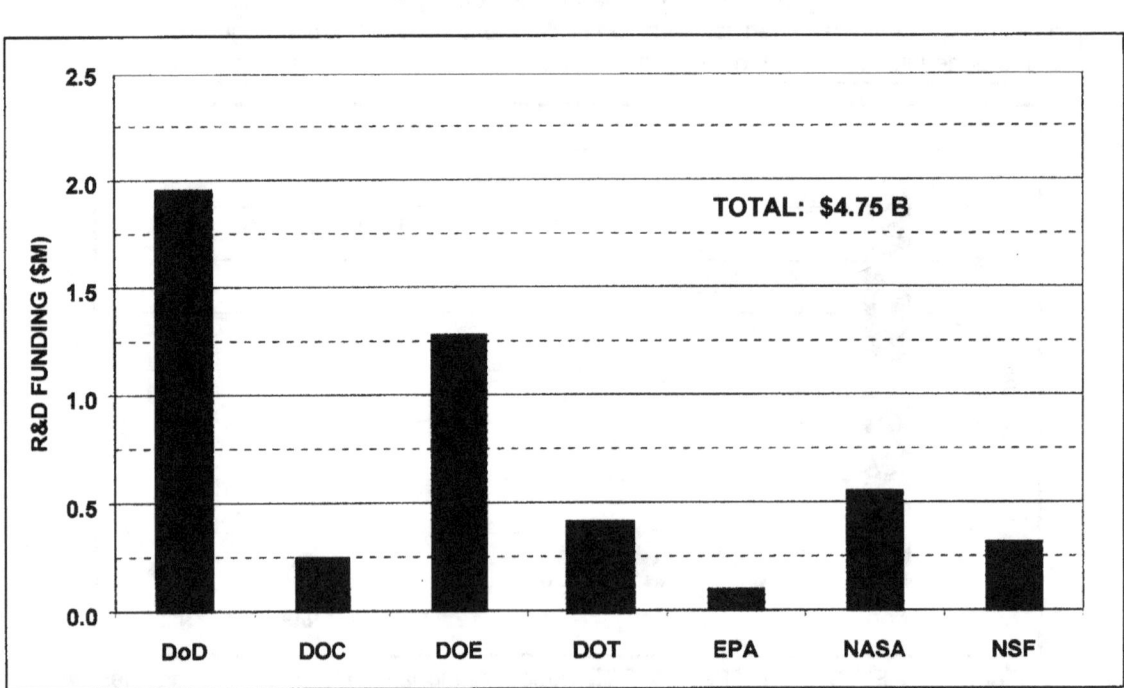

Figure 2-3. FY1998 Transportation-Related Federal R&D Budget Authorization for Selected Agencies.

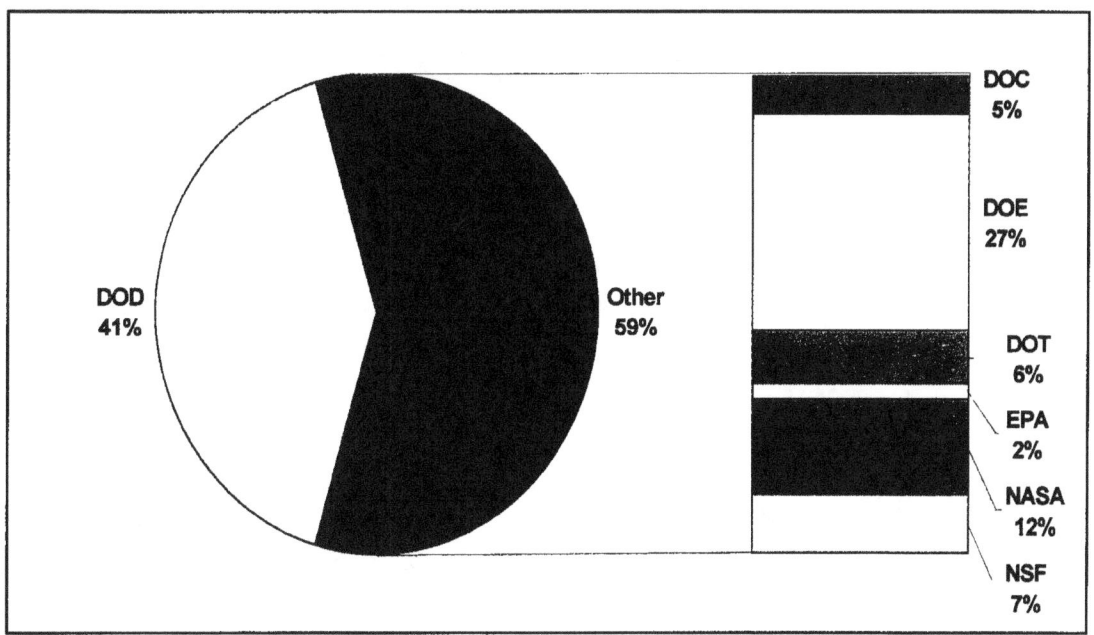

Figure 2-4. FY1998 Transportation-Related R&D Budget Authorization for Selected Agencies

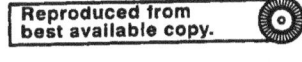

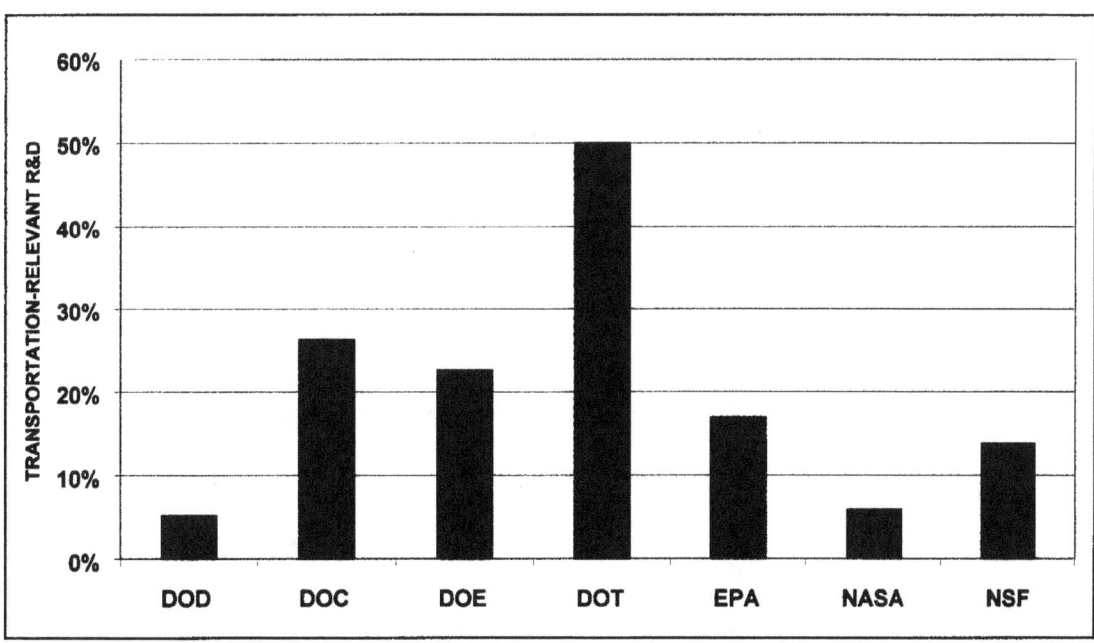

Figure 2-5. Percentage of FY1998 Agency R&D Budget Authorization Identified with Transportation-Related Enabling Research.

3. "BREAKTHOUGH" RESEARCH DIRECTIONS

BREAKTHROUGH RESEARCH

Enabling research such as that described in this plan is an essential element in maintaining the continuous evolutionary advances needed to achieve a transportation system that can meet the needs of the 21st century. However, recent decades have made clear that success in the future will also require more than evolution—it will also rest on periodic injection of breakthroughs from the world of basic research that significantly alter and expand the technological options available to the transportation enterprise.

It is the nature of basic research that its practical implications are not immediately clear. Often, many years go by before the major impacts are felt. However, the technological acceleration in the latter part of the 20th century has generally shortened the period between scientific findings and invention and realization of real-world innovation. Potential users as well as the scientific community are becoming more and more sensitized to the potential value of research products, so that the interval between discovery of new knowledge and practical application has become relatively brief.

Further, the pressure of the challenges we face, and the rewards for successful exploitation of scientific and technical advances, often combine to encourage research in areas where the outputs are particularly likely to have special value. This is as true for transportation as for any other human endeavor. For that reason, this section of the *National Strategic Transportation Research Plan* identifies three of the numerous areas of important basic research now being pursued which have especially clear and substantial potential benefits for the designers, builders, operators and users of future elements of the national and global transportation system. These areas are:

- Nanotechnology
- Biofuels
- Complex Systems and High-Confidence Software

All three areas are addressed in the specific enabling research discussion in Section 4 of this plan, but their importance warrants the more focused discussion that follows.

NANOTECHNOLOGY

"Nanotechnology" is the building of devices and materials at the level of atoms and molecules and the exploitation of the novel properties at this scale. It gets its name from "nanometer," the unit of measurement representing one-billionth of a meter, or about ten times the size of an individual atom. (In comparison, microtechnology is at the level of microns—a millionth of a meter—gargantuan by the standards of nanotechnology.) "Nanotechnology" generally refers to work done at the scale between 0.1 and 100 nanometers. The novel properties are due to the different behavior of things at this scale as compared with either isolated molecules or larger structures.

Nanotechnology implies the direct control of atoms and molecules. Broadly, it comprises (1) the design of atomically engineered "building blocks"; (2) the assembly of these building blocks into new, "nanostructured" materials with specific characteristics; and (3) the assembly of these materials into useful devices.

There are two overall approaches to building things at the nanoscale. The first is to etch, chisel, or sculpt such features into an existing, larger structure, using techniques such as scanning tunneling or atomic-force microscopy or various forms of lithography (the process now used to make computer chips). Nano-size gears and smaller integrated circuits are just some of the objects being fabricated using this approach, which is sometimes referred to as "top–down." The second method—and the most revolutionary—is the "bottom–up" approach: building things up from the atoms and molecules themselves. One recent breakthrough resulting from bottom–up assembly is the single-electron transistor.

Yet, although researchers can now make single molecular structures in the laboratory, they have yet to find a cheap and commercially feasible way of mass-producing them. One answer may lie in a process called "self-assembly." This relies on chemistry to position atoms and molecules, taking advantage of certain molecules' abilities to arrange themselves in complex structures.

Impact on Transportation

In the last few years, nanotechnology has yielded products with sales totaling billions of dollars, including giant magneto-resistance multilayers (for computer memory); nanostructured coatings (for data storage and the photographic industry); nanoparticles (drug-delivery devices in pharmaceuticals and colorants in printing); and nanostructured materials (nanocomposites and nanophase metals). It is revolutionizing virtually every area of technology and will have a direct and dramatic impact on transportation.

Nanotechnology's potential benefits for transportation are broad and pervasive: lighter, more efficient cars using nanostructured materials; corrosion-free bridges and no-maintenance roads; tiny "traps" that remove pollutants from vehicle emissions; and robotic spacecraft that can explore the solar system and yet weigh only a few pounds. Among the potential transportation breakthroughs are the following:

Information Technology. With molecular electronics, a single computer chip could hold billions of miniscule transistors, making computers orders of magnitude more powerful than they are today. Specific applications for transportation include: (1) uninhabited vehicles for civilian and defense use; (2) advanced communications that maximize the benefits of intelligent transportation systems and obviate the need for some travel altogether; and (3) advanced sensors that continuously monitor the condition and performance of infrastructure, vehicles, and operators.

Materials. Nanotechnology will yield transportation materials that are lighter, stronger, and, ultimately, programmable—reducing costs through longer service life and lower failure rates. Among the key applications are: (1) nanocoating of metallic surfaces to achieve super-hardening, low friction, and enhanced corrosion protection; (2) "tailored" materials for infrastructure and vehicles; and (3) "smart" materials that monitor and assess their own status and health and repair any defects—including self-healing, fire-resistant materials in vehicles and aircraft.

Aeronautics and Space. New materials developed through nanotechnology will meet the strength, weight, and thermal stability requirements of space planes, rockets, space stations, and high-speed aircraft. Moreover, nanotechnology will permit the ultra-miniaturization of space systems and equipment, including the development of smart, compact sensors; miniscule probes; and microspacecraft. Applications include: (1) economical supersonic aircraft; (2) low-power, radiation-hard computing systems for autonomous space vehicles; and (3) advanced aircraft avionics.

Environment and Energy. Nanotechnology has the potential to reduce transportation energy use and its impacts on the environment. For example, nanosensors could be used to monitor vehicle emissions and to trap any pollutants. Other applications include: (1) nanoparticle-reinforced materials that replace metallic components in cars—those now being developed could reduce CO_2 emissions by more than five billion kilograms a year; (2) replacement of carbon black in tires with nanoparticles of inorganic clays and polymers, leading to tires that are environmentally friendly and wear-resistant; and (3) carbon-based nanostructures that serve as "hydrogen supersponges" in vehicle fuel cells.

BIOFUELS

The transportation sector is currently heavily reliant on petroleum-based fuels such as gasoline and diesel fuel. Petroleum is not renewable on the time scale during which it is used, and is becoming increasingly obtained from politically unstable regions. This creates the potential for both long-term price increases and short-term supply disruptions. Petroleum also consists mainly of carbon that is oxidized to form carbon dioxide—the dominant human-induced greenhouse gas—during the combustion process.

Many other possible fuel feedstocks have the potential to increase energy security and/or reduce net greenhouse gas emissions. Examples include coal, tar sands, shale, and natural gas Another alternative is electricity, which can be used to either recharge electric vehicles or produce hydrogen for use in internal combustion engines or fuel cell vehicles. Electricity can be generated using a wide range of energy sources, including renewable sources like wind and solar power. Another increasingly attractive potential option is based on the use of biomass. Biomass offers the unique promise of addressing both energy and climate concerns in a manner that could be reasonably compatible with the existing infrastructure for distribution and delivery of liquid transportation fuels.

Biomass, which is defined as plant matter of recent origin, represents a massive renewable resource. In the US, from 10 to 40 billion gallons of gasoline could be displaced with biomass ethanol through the use of wastewood, agricultural waste, cropland, and rangeland/grassland. Carbon dioxide from the atmosphere and water from the earth are combined in the photosynthetic process to produce carbohydrates (sugars) that form the building blocks of biomass. The solar energy that drives photosynthesis is stored in the chemical bonds of the structural components of biomass. When biomass is burned efficiently, oxygen from the atmosphere combines with the carbon in plants to produce carbon dioxide and water. The process is cyclic because the carbon dioxide is then available to produce new biomass.

The chemical composition of biomass varies among species, but biomass typically consists of about 25 percent lignin and 75 percent carbohydrates or sugars. The carbohydrate fraction consists of many sugar molecules linked together in long chains or polymers. Two larger carbohydrate categories that have significant value are cellulose and hemi-cellulose. The lignin fraction consists of non-sugar type molecules linked together in large two-dimensional sheet-like structures that look like "chicken wire."

In part because of process inefficiencies, fuels from biomass are not economically competitive at this time. Breakthrough in biological and thermal conversion techniques could make it possible to transform biomass into fuels much more efficiently. Although serious ethical questions and potential risks to ecosystems must be considered, genetic engineering could be the basis for a breakthrough in biological processing of biomass. Separately, new thermal conversion techniques coupled with chemical catalysis could make it possible to exploit the previously discarded lignin fraction by converting it into valuable chemicals that are currently derived from non-renewable fossil sources.

Although land availability limits the potential of biomass to supply all of the energy required for transportation, breakthroughs in the areas mentioned above could make it possible to meet a significant fraction of that demand in a manner that also significantly reduces greenhouse gas emissions.

Impact on Transportation

Concern over the impact of carbon dioxide and other greenhouse gases on the global climate have been increasingly steadily in recent years. The transportation sector is a key player in this complex issue, producing approximately 1/3 of total human-produced carbon dioxide emissions. (For example, combustion of a single gallon of gasoline generates about 20 pounds of CO_2.) Should the US and other national governments at some point find it necessary not only to prevent further growth in emission of greenhouse gases, but also move to achieve significant reductions, the impact on transportation—and the associated consequences for economies the world over—would be very substantial. There would not only be a reduction in the level of transport of goods and people, but it would inevitably become much more costly. Allocation of the pain of any such transition would be a highly contentious matter, and the choices made could generate serious strains on the social fabric.

Under those circumstances, availability of fuel feedstocks that play a far more benign role in CO_2 production would be enormously valuable to the whole world, and particularly to the US, which accounts for the largest petroleum consumption of any nation. This consideration alone warrants continued aggressive research to find practical ways to exploit biomass-based fuels.

COMPLEX SYSTEMS AND HIGH-CONFIDENCE SOFTWARE

The rapid development and interrelationship of computer, sensing, communication and navigation technologies is having a powerful impact across the spectrum of transportation operations in all modes. Originally nurtured by space- and defense-related R&D, and now driven largely by global business and consumer markets, this revolution is proceeding rapidly in the private sector. Public sector applications, particularly in transportation, are for the most part only in early stages of implementation, but are seen as having great near- and long-term promise. One important and highly visible area currently being actively addressed is applications to traffic management on ground, air and sea, where congestion often exacts a high societal cost in terms of time, accidents, air pollution and quality of life for all those affected by it. More generally, the efficiency of transportation system operations, regardless of the mode, is also critical to the productivity and overall health of the entire economy.

Progress is already rapid in development of systems in which information technologies and ready access to many types of data are integrated into virtually all system elements and functions to enable greater efficiency and improved performance. The rapid infusion into the marketplace of "smart" products, which combine sensing and computation to make a wide range of decisions for their users, is just the first wave of ubiquitous automation. However, as their power and economic attractiveness stimulates large-scale application of the next generations of intelligent transportation systems, severe challenges can be expected. In particular, as safety-critical systems (e.g., motor vehicle crash avoidance, aircraft flight management technology) and very widely distributed systems (e.g., highway traffic control across an urban area, or global air traffic management) proliferate, the reliability and failure modes of the technology will become ever-more critical. At the same time, system complexity and sophistication will similarly be increasing, to the degree that a truly comprehensive analysis of reliability and failures under all circumstances becomes extremely difficult.

Thus, the steadily increasing complexity of intelligent or otherwise highly-automated systems, accompanied by ever-greater dependence on them, will require new extremes of reliability, robustness, and non-catastrophic failure modes ("graceful degradation") in the face of subsystem failures or severe and unexpected circumstances. The challenge thus posed is daunting: not only must the design process generate systems meeting these standards, but means must also be developed to confirm that the resulting systems meet the goals. As system size and complexity grow, and the systems themselves involve large populations of distributed and often autonomous users, new conceptual approaches and development tools will be required. Transportation, in which safety considerations must always be paramount, is a key user of technological advances in this area, but the necessary enabling research is most naturally done in the information technology community, both public and private. The Departments of Defense and Commerce have generally had the lead in this discipline.

Impact on Transportation

If the challenge of developing super-reliable and highly forgiving or self-correcting software-based systems is successfully met, the safety and performance of the transportation system will benefit greatly. Advances that enhance operator situational awareness and capabilities, and correctly identify and respond to hazardous circumstances, could provide the key ingredient needed to bring about a significant decline in transportation accident and fatality rates, which have been relatively flat in recent years.

Systems operating at or near full capacity can provide very high efficiency, but this is coupled with severe consequences from any disruption, such as weather constraints affecting busy air routes and accidents or breakdowns on crowded highways. The fullest use of automation for traffic management in all modes will only be possible when very high capacity automated systems achieve sufficiently high reliability and robustness to avoid the risk of new extremes of gridlock and system collapse should any element fail or the operating environment alter.

4. AREAS OF ENABLING RESEARCH

OVERVIEW

Innovations in transportation generally result from the application of a wide range of scientific and engineering disciplines, including many that do not have a specific transportation focus. A solid foundation for developing the transportation technology needed for the 21st Century must include research in a broad spectrum of topics. The long-term and often diffuse benefits of wide-ranging research are often such that market forces may be insufficient to motivate private investment. It is in these areas that enabling research performed or funded by the Federal government can pave the way for future technological advances from the private sector.

This plan uses the seven long-term research areas defined in the NSTC *National Transportation Science and Technology Strategy* as a structure for analyzing Federal transportation-related enabling research. They are:

- Human Performance and Behavior;
- Advanced Materials and Structures;
- Computer, Information, and Communication Systems;
- Energy, Propulsion, and Environmental Engineering;
- Sensing and Measurement;
- Analysis, Modeling, Design, and Construction Tools; and
- Social and Economic Policy Issues

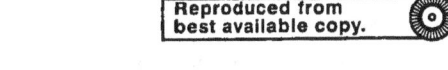

These research areas provide an insight into the existing and the potential commitment within the Federal community. Figure 4-1 shows the breakdown of Federal transportation-related research funding in terms of each of the seven enabling research categories described above. *Computer, Information, and Communications Systems* and *Energy, Propulsion, and Environmental Engineering* each comprise 29 percent of transportation-related research. *Advanced Materials and Structures* is the other dominant category, accounting for almost one-quarter of the total. The remaining four research areas receive a substantially smaller fraction.

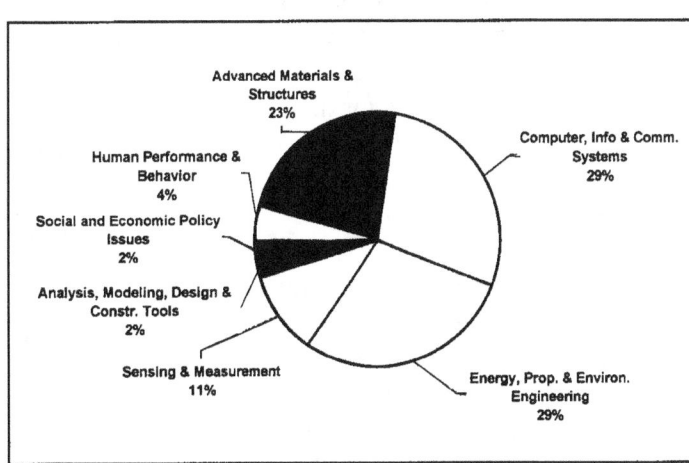

Figure 4-1. FY1998 Total Transportation-Related R&D Budget Authorization by Category

As shown in the previous section, the Department of Defense, with its extensive technology requirements and long tradition of supporting basic and applied research, dominates the distribution shown in Figure 4-1. It is therefore useful to examine the DoD and civil agencies separately. Figure 4-2 shows the distribution of DoD transportation-related research activities, in terms of enabling research categories, for fiscal year 1997. Within the DoD, *Computer, Information, and Communications Systems* is by far the largest research area, accounting for 49 percent of the agency's transportation-related research budget.

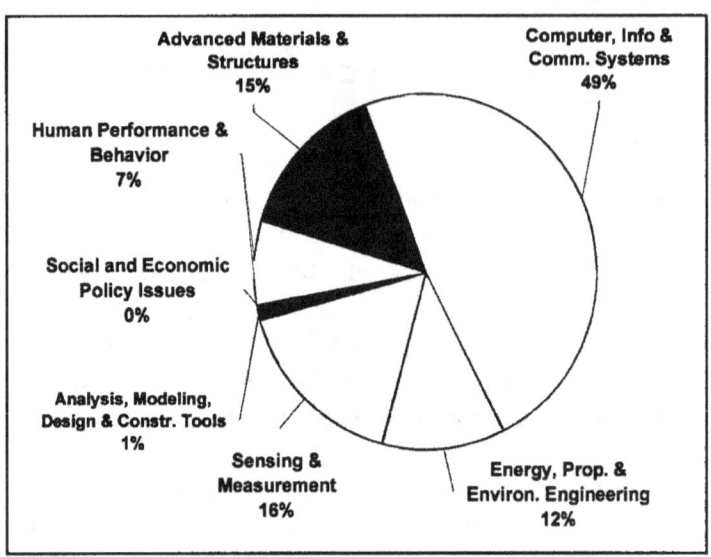

Figure 4-2. FY1998 Transportation-Related R&D Budget Authorizations Categorized by Area of Enabling Research (Dept. of Defense only).

The picture is significantly different for the non-DoD agencies indicated in Figure 4-3. Three categories account for 84 percent of their total transportation-related R&D: energy (due largely to the substantial DOE transportation energy program), materials (predominantly NASA and NSF), and computer and communication technologies (principally DOT and NSF).

Figures 4-4 and 4-5 summarize transportation-related research budgets, by agency and enabling research category. Figure 4-4 presents the data in terms of DoD and the aggregated totals for the civil agencies; Figure 4-5 provides the detailed allocations among each of the civil organizations.

A description of each enabling research category is provided on the following pages. It includes the primary agencies active in the category, a discussion of its relevance to transportation, principal types of application-oriented R&D supported by the enabling research, and representative current examples of enabling research.

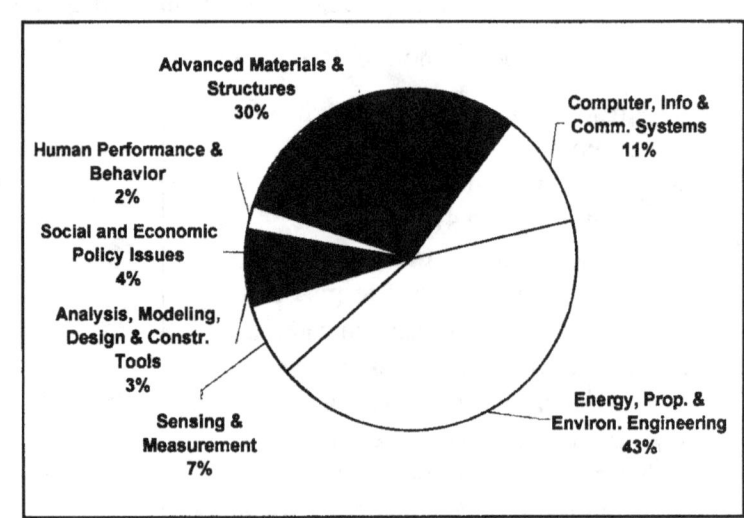

Figure 4-3. FY1998 Transportation-Related R&D Budget Authorization for Selected Agencies other than Dept. of Defense Categorized by Area of Enabling Research.

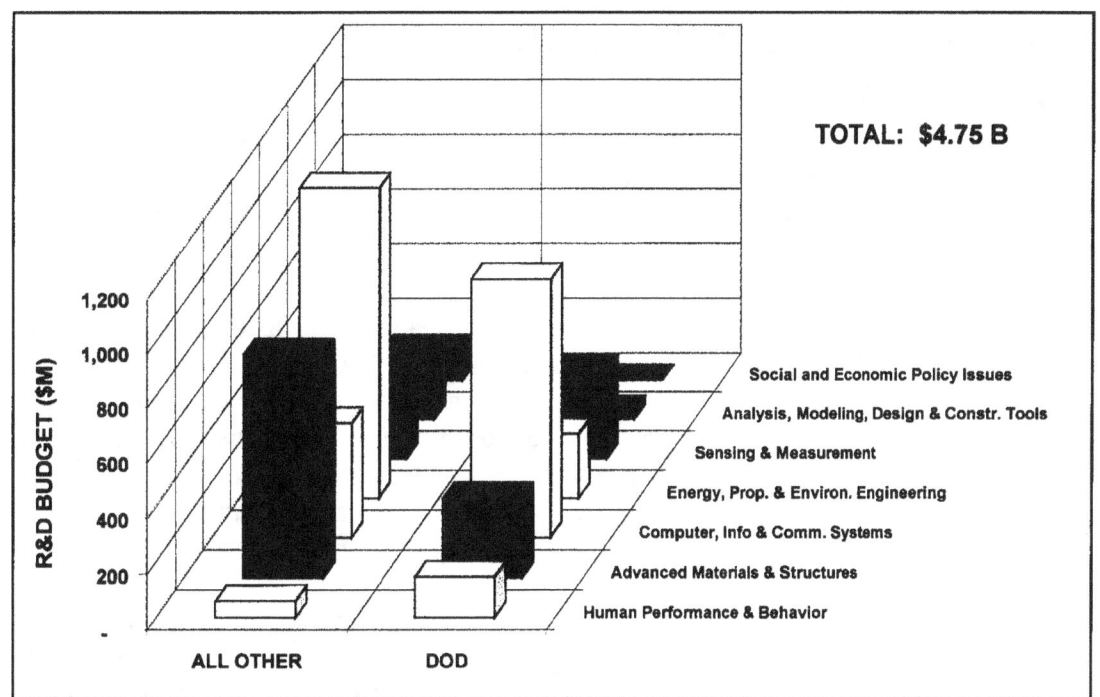

Figure 4-4. FY1998 Transportation-Related R&D Budget Authorization Categorized by Area of Enabling Research.

Reproduced from best available copy.

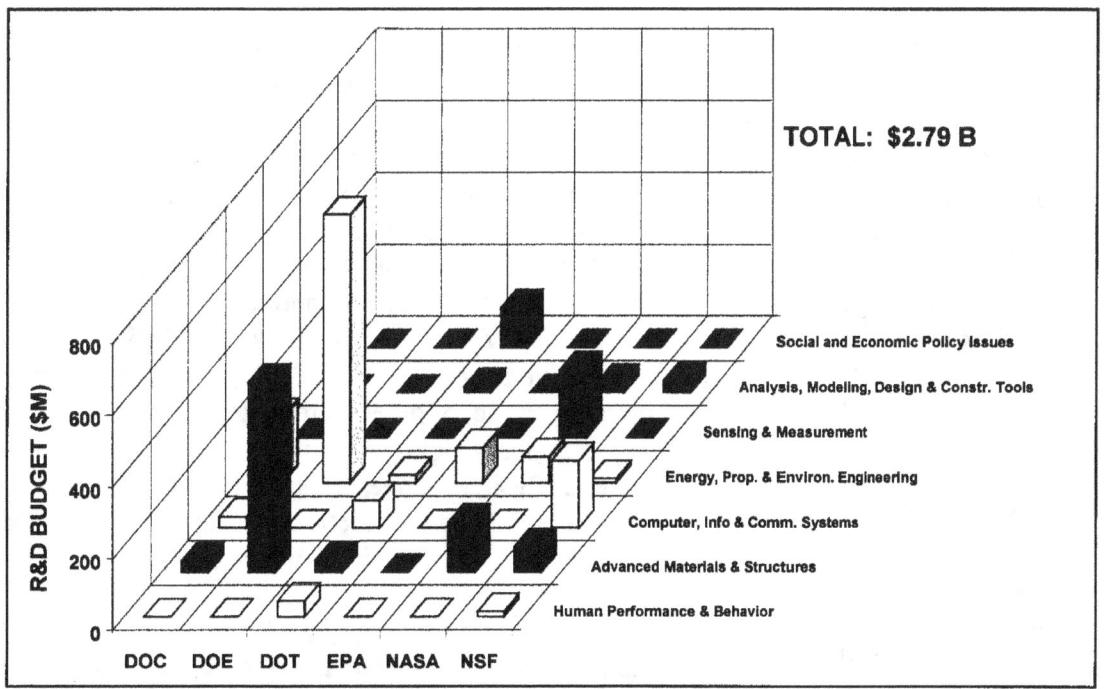

Figure 4-5. FY1998 Transportation-Related R&D Budget Authority By Agency and Area of Enabling Research for Agencies other than DoD.

HUMAN PERFORMANCE AND BEHAVIOR

Overview

Transportation projects seek to achieve maximum performance at a minimum cost. To achieve these goals, systems and technologies must be designed and implemented based on an understanding of user needs, as well as the many factors affecting the ways that people interact with automated systems.

Representative Applications:
- **Collision Avoidance Systems (DOT/NHTSA)**
- **Development and Application of Technology to Monitor Driver Fitness (DOT/FMCSA)**
- **Fatigue Research, (DOT-USCG/FMCSA)**
- **Human Factors in Air Traffic Control (DOT/FAA)**
- **Railroad Dispatcher Stress and Fatigue Studies (DOT/FRA)**
- **Intelligent Vehicle Initiative (DOT)**
- **Flight Deck Human Factors (DOT/FAA, NASA)**
- **Design to Accommodate Aging Drivers (DOT/FHWA, DOT/NHTSA)**
- **Human Cognition and Perception, Social Psychology, Decision, Risk, and Management (NSF)**
- **Implementation of Piloting Navigation Aids (DOT/MARAD)**
- **Human Systems Research into Information Display/Performance Enhancement; Design Integration; Personnel Performance and Training (DoD)**

Human error is a leading contributor to transportation-related safety problems; it also may create operational inefficiencies that reduce overall productivity. To some extent, these problems occur because basic system design, operational procedures, and training programs fail to consider the human performance limitations of transportation system users and operators.

Behavioral science research provides a critical foundation for designing systems and procedures that will be effective under real-world conditions and will apply to multiple modes of transportation. In particular, system control and operations will be improved through better understanding of the human response to the environment.

There are several key areas of transportation applications related to human performance and behavior. At the most fundamental level, research efforts are in progress or are proposed that are aimed at understanding how transportation system users and operators perceive, process, and act upon information in "real world" situations. For example, certain physical design characteristics of roadways and intersections may contribute to driver error, by making it difficult for drivers to perceive the need to slow down, their distance from merging traffic, or other factors. Roadway design issues are particularly significant for younger, less experienced drivers. Design changes, ranging from different striping patterns to altered intersection and curve geometry, can help reduce driver error and enhance roadway safety and efficiency.

Another critical area is understanding the factors that contribute to fatigue, particularly among operators of vehicles such as large ships—where there is minimal room for error—and the development of fatigue countermeasures. Fatigue countermeasures also are crucial for transportation operations personnel such as air traffic controllers, who are responsible for the safety of numerous transportation system users and operators.

A third area develops automated systems that provide active safety enhancement by augmenting human capabilities. They provide early warnings of potential danger (e.g., one vehicle following another too closely) and, in some cases, can take over some degree of vehicle control to help avoid collision (e.g., applying brakes). Understanding the ways in which vehicle operators process and act upon such warnings is a necessity if the systems are to be effective.

In certain high-risk situations, such as marine navigation, human error often is attributable to inaccurate or misunderstood communication. To help remedy this problem, important research is proposed to enhance the communication of transportation systems operations teams, such as marine pilots and vessel crews.

Illustrative Current Enabling Research

The NSF Human Cognition and Perception program supports research on human perceptual and cognitive processes. Topics addressed include vision, audition, perception, attention, object recognition, language processing, spatial representation, motor control, memory, reasoning, and concept formation.

Other identified leading edge research areas come from DOT's Human Factors Coordinating Committee. Two initiatives on human factors, one, dealing with advanced institutional technology, and the other with alertness and fatigue, have evolved from this Committee's work. These initiatives are multi-modal and multi-agency.

The Defense Department has long recognized the criticality of human behavior and performance aspects of system design and conduct of operations.[5] Although military situations and circumstances are very different from those that characterize the transportation system, the underlying principles that guide design of equipment and systems, training programs and operational procedures are basically the same. The DoD research program in this area is therefore highly relevant to application in transportation.[6]

Defense design integration and supportability goals include: (1) developing a national technology base in human performance modeling and assessment; (2) designing tools for physical accommodation; (3) devising methods for human error assessment; (4) creating tools for estimating and evaluating human performance requirements; (5) demonstrating how to achieve effective crew system integration during design; and (6) building tools to streamline and enhance the infrastructure system. The ultimate aim is to improve system effectiveness, availability, and affordability. All of the design integration tools are set in the context of weapon system engineering, but have clear application to transportation.

---

[5] The following material is drawn from the DoD Internet site at https://ca.dtic.mil/dstp.
[6] Ibid.

In the area of information display and operator performance, DoD goals are (1) to enhance situation awareness through exploitation and integration of emerging sensor, display, and processor technologies for organizing, managing, and displaying vast amounts of information; and (2) to greatly enhance mental performance while adapting emerging display and performance technologies to unique tactical requirements. These complementary goals will significantly enhance performance by optimizing the utility of the information and the ability of the operator to absorb and act upon it.

Specific targets for more than five years in the future include: (1) advanced data fusion and processing to provide near-real-time information for operational awareness and decision-making; (2) large color graphic displays to support 3-D views of tactical information; (3) 3-D audio, speech recognition, and color helmet displays to support threat warnings; (4) architectures to significantly augment cognitive, perceptual, and physical task performance applicable to all services, defense agencies, and private-sector spin-offs; and (5) immersive virtual reality devices to support crew stations including remotely piloted vehicles.

Promising Areas for Interagency and Public-Private Partnerships

Advanced Instructional Technology: Even with a greatly improved understanding of human performance, training and education are critical for ensuring that vehicle operators are knowledgeable about their vehicles' safety systems and safe operational procedures. Research in this area would support the development of interactive programs (such as CD-ROM video techniques and simulation) to train and evaluate operators under a wider range of operational scenarios than is currently possible.

Enhanced Alertness and Work Readiness: Research is needed to study and quantify the effects of fatigue, work–sleep cycles, working environment and culture, boredom, and drug and alcohol use; response to emergency situations; testing of readiness to perform duties; and interactions among co-workers.

Humans and Automated Systems: Critical information for designing systems and technology that enhance, rather than hamper, performance would offer real benefits to transportation. Examples of topics in this area include human cognition and perception, reaction times, and visual acuity. Research partnerships in this area would examine the properties and performance of materials and technologies with physical infrastructure applications.

ADVANCED MATERIALS AND STRUCTURES

Overview

Technical advances in the defense and consumer sectors have produced a wide variety of new materials and techniques. This rich inventory includes high-performance concrete, new steel alloys, innovative composite materials and adhesives, imaginative structural concepts, computer-aided design techniques, automated construction and maintenance tools, and new approaches to corrosion protection and control. All of the agencies considered here except EPA conduct research in this area, with DoD, NASA and NSF as the dominant players. Enabling research in this area supports the application of these advances to both transportation infrastructure and vehicles. Such applications may include demonstrations of their effectiveness, long-term viability, and cost-competitiveness in enhancing safety and performance. Under the DoD Technology Reinvestment Project, research has been conducted on low-cost processing of specialty metals and ceramic materials applications, as well as electric and hybrid vehicles.

Representative Applications:
- **Technology Reinvestment Program (DoD)**
- **High-Speed Research Program (NASA)**
- **Advanced Space Propulsion Technologies (NASA)**
- **Shipbuilding Research, Maritime Administration (DoD, DOT/MARAD)**
- **Design and Crash Testing of Composite Automobiles (DoD)**
- **High Performance Materials for Infrastructure and Structural Applications (DOT/FHWA)**
- **Advanced Materials Processing Program, Nano-Fabrication Users Network (NSF)**
- **Materials program for light and heavy highway vehicles to reduce weight and improve fuel economy (DOE)**

The materials used to manufacture aircraft, ships, and surface vehicles also have a significant impact on transportation safety and operating performance, as well as on energy use. Each potential application of an innovative material poses new challenges with respect to material costs, manufacturing processes, failure mechanisms, environmental concerns, and cost compatibility with transportation uses.

Some particularly innovative approaches are being considered. For example, improvements in night vision of drivers and the conspicuity of pedestrians at night are being sought through use of ultraviolet light and specialized fluorescent materials.

Another promising direction involves use of advanced materials in conventional applications in order to improve performance. For example, research is underway to evaluate the feasibility of using composite highway guardrails in place of steel.

Improvement of the performance and/or cost effectiveness of existing materials is also the subject of current R&D. Work underway in this area includes studies of high-performance concrete, steel, and wood construction materials.

Also important is research focused not only on the development of new materials, but also on the required changes in existing manufacturing and production processes in order to use these new materials. Research also is proposed to investigate the costs and potential problems associated with using certain new materials.

Illustrative Current Enabling Research

The NSF supports research directed toward increasing the knowledge base in the areas of construction, geotechnology, structures, dynamics and control, mechanics, and materials, as well as the reduction of risks induced by earthquakes and other natural and technological hazards.[7] The Construction, Geotechnology and Structures program emphasizes new discoveries in the design, construction, maintenance, and operation of facilities that are safe, long lasting, efficient, environmentally acceptable, and economical. Increased understanding is sought concerning advanced polymer materials, high performance steel and concrete materials, deterioration of construction materials, as well as safety and reliability of bridges.

Additional research activities will increase the current understanding of the science and technology used to design, analyze, diagnose, repair, remediate, retrofit, and enhance the performance of constructed facilities. Insight is also needed into the interactions between natural and constructed environments. The knowledge gained will improve the management and performance of new and existing infrastructure systems.

The Department of Commerce carries out research to develop high-performance construction materials with superior mechanical and durability properties.[8] However, widespread use is hindered by a lack of understanding about the structural performance of these materials and the lack of design standards. The objective of current research is to provide the basis for design criteria and evaluation methods for high-performance materials. Specific projects include: shear strength of high-strength concrete beams, curing of high-performance concrete, performance of high-strength concrete during fire, and use of fiber-reinforced polymers for rehabilitation and repair of structures. Related R&D performs analytical, laboratory and field research. It relates to the development of methods that measure and predict service life of construction materials. The development of technical bases for improving criteria and standards are used to evaluate, select, use, and maintain construction materials. They also improve tools that make decisions in selecting construction materials, including high performance concrete and steels.

In the Department of Defense, pavement design, repair, and material criteria are under development to ensure reliable support for current and future-generation aircraft and vehicles used in military operations.[9] Relevant innovation requirements include advanced material characteristics and construction technologies, advanced analytical models to allow rapid and accurate pavement capacity determinations, and criteria for local materials use. This minimizes logistical construction burdens and reduces time constraints. The DoD civil engineering program uses and contributes to the Civil Engineering Research Foundation, a coordinated industry and government 10-year plan for high-performance construction materials. It also builds on fundamental advances by the NSF's university-executed research in several engineering disciplines and technologies (cementious materials, composites, nondestructive testing, structural and geotechnical dynamics).

---

[7] National Science Foundation, *Guide to Programs, 1998*
[8] See Internet site at www.nist.gov.
[9] This discussion is based on material from the DoD Internet site at https://ca.dtic.mil/dstp.

Relevant basic research by DoD focuses on enhancing understanding of stress-strain relationships at the smallest aggregate of particles within the soil matrix, the constitutive behavior of construction materials, and soil-moisture-strength relationships as a function of climatic influences across the world. These efforts will directly contribute to development of a high-resolution, high-fidelity mobility model that accurately predicts worldwide vehicle movements, both on and off the road. In addition, research on constitutive behavior and micromodeling of asphalt concrete provides basic understanding of asphalt response to loads. Basic research on the constitutive behavior of concretes is leading to improved predictions on the responses to concrete slabs, to projectile impact and penetration, and to concrete structural elements in blast loadings.

The DoD program directed toward life-extension capabilities for the Navy's waterfront infrastructure will increase its load capacity to meet new mission performance and safety requirements. This will be demonstrated through: (1) structural composite reinforcements to accommodate high concentrated crane loads for which the structural design and the old piers were not designed; (2) concrete repair technology with extended durability (from the present three years to 15 years); (3) corrosion stabilization steel reinforcements using plasma-sprayed titanium sacrificial electrodes to extend pier life by 20 to 30 years; and (4) improved modeling for assessing pier structural safety with respect to operational loads and recently updated environmental resistance requirements.

Promising Areas for Interagency and Public-Private Partnerships

Advanced Infrastructure Materials: Research partnerships in this area would examine the properties and performance of materials and technologies with physical infrastructure applications. Examples include high-performance concrete, new steel alloys, innovative composite materials and adhesives, imaginative structural concepts, computer-aided design techniques, and automated construction and maintenance tools.

Materials for Transportation Vehicles: This research addresses materials with potential use in the manufacture of aircraft, ships, and surface transportation vehicles; for example, the use in body structures of high-performance steel, aluminum, magnesium, and glass- and carbon-fiber composites. Such research partnerships would need to consider material cost, manufacturing processes, failure mechanisms, and environmental characteristics.

Advanced Manufacturing and Construction: This research would support the development of advanced technologies for infrastructure construction and materials manufacturing. A major focus would be work leading to technologies that improve the sustainability of materials production by reducing waste, pollution, and emissions generated in the manufacturing process.

COMPUTER, INFORMATION, AND COMMUNICATIONS SYSTEMS
Overview

Modern transportation systems require continual exchange, processing and use of accurate, timely information. As information infrastructures are overlaid onto the physical transportation infrastructure, ready access to information is becoming integrated into virtually all system elements and functions. Taking full advantage of these innovations to improve efficiency, safety, and performance requires research and development focused on system concepts, and on the characterization of alternative configurations and technical choices.

In addition, computer and communications systems often cannot exchange data quickly or accurately because transportation systems structure and handle information in different ways. Basic interoperability standards for the electronic interchange of transportation system data are being addressed by trade and technical organizations. Nonetheless, the Federal government has a critical role to play in helping to develop uniform standards for transportation-related computer, information, and communication systems.

Representative Applications:
- **Intelligent Transportation Systems (ITS) (DOT)**
- **Identification and Evaluation of Alternative Communication Systems for ITS (DOT/FHWA)**
- **New Traffic Control Data Communication Systems for ITS (DOT/FHWA)**
- **Positive Train Control Systems (DOT/FRA and DOT/FTA)**
- **Port Operations Information for Safety and Efficiency (DOT/USCG)**
- **Information Technology for Improved Aviation Operational Systems (NASA)**
- **Knowledge and Distributed Intelligence High-Performance Computing and Comm. (NSF)**
- **Next Generation Internet Initiative (Multiple Agencies)**
- **High Confidence Systems R&D (Multiple Agencies)**
- **Marine Virtual Data Hub (DOT/USCG)**

The growing complexity of intelligent systems and transportation's dependence on them requires the systems to be highly reliable and robust; it also renders them increasingly vulnerable to accidental or deliberate service interruptions. Federal research is conducted on systems development, modeling, and verification techniques to provide users with high levels of security, reliability, and restorability. Systems that employ these techniques will mitigate component failure and malicious manipulation and will take corrective action in response to perceived threats or actual damage.

Important applications currently underway include development of standards and protocols for communication and data exchange among traffic control devices. These may involve different types and different manufacturers. In addition, ongoing projects are focused on issues related to the operation of transportation-related communication systems within the overall communications environment. For example, exploration of this area is intended to resolve technical issues associated with the allocation of specific radio frequency spectrum allocation for ITS applications.

Another promising area is the application of advanced computational and modeling techniques to traffic simulation. For example, one current project uses emerging theories of human behavior to develop simulation models of vehicle actions in different situations.

Data fusion is a particularly promising discipline in the collection and dissemination of information vital to the operation of transportation systems. For example, many parties are exploring enhanced means to gather, process, and disseminate weather information to highway maintenance/operations units and travelers.

Illustrative Current Enabling Research

In many ways, the information environments for the industrial, commercial, and financial communities mirror the military information environment. Both the defense establishment and multinational corporations and financial institutions have the need for global data access. Movement of global markets requires very rapid response to change and guaranteed availability. This creates the need for similar distributed information environments that provide location-transparent access to globally distributed data.

As a result, the technical issues in this category of enabling research show particularly rich commonalities and synergies between the needs of the defense and civil sectors. Global corporations and financial institutions have the need for global data access. Their multinational status requires support for heterogeneity. Movement of global markets requires very rapid response to change and guaranteed availability. This creates the need for similarly distributed information environments that provide location-transparent access to globally distributed data.

Defense-Related Enabling Research

For DoD, the execution of critical real-time decisions can shape the outcome of battles. The uncertainty or lack of availability of information conspires to slow and confuse the process. There is a clear need for decision aids that permit the rapid assessment, planning, and execution of missions to ensure swift attainment of goals through constraint-based, information-intensive systems. These decision-making systems must organize, explore, and recommend options across a spectrum of operations.[10]

Accordingly, the Defense Department gives major emphasis to acquiring, organizing, and manipulating information needed to accomplish military missions as well as to manage all its operations, in peace and time of conflict, efficiently and expeditiously. One of the primary objectives of R&D in this area is to achieve information superiority by meeting the need for a flexible information-presentation system that can be configured rapidly, and a structure dynamically adapted to optimize operations. This research applies leading-edge computing and software technologies to improve performance significantly. It eliminates laborious, time-consuming manual procedures and processes that pervade operational task assessment, planning, and execution. Computer-aided processes and procedures replace exclusively human ones.

The major challenges addressed by this research are as follows:

---

[10] The following discussion is based on material from the DoD Internet site at https://ca.dtic.mil/dstp.

- Develop applications that organize and effectively present complex, distributed information. (These use advanced pattern recognition algorithms, knowledge bases, and goal-directed and constraint-based reasoning that employ intelligent agents for semiautomated, intelligent information retrieval, fusion, and presentation.)
- Fuse planning information with actual information in real time.
- Provide real-time simulation, collaborative planning, and rehearsal with sufficient fidelity on tactical systems to influence mission outcomes.
- Create decision support in the presence of uncertain, incomplete, or absent information.
- Build applications for dynamic scheduling and coordination of assets for interdependent tasks.
- Use collaboration tools that permit the spectrum of operations to be performed by remote, dispersed elements of a task force.

Many critical information technology challenges being addressed by DoD are centered in three areas. The first is the infrastructure for distributed environments. The second relates to mechanisms supporting information services management that reside within the distributed environment. The third is the ability to deploy assured information services.

In the first area, the critical technical challenges are: (1) scalability to several thousand nodes and schedulability of time-critical operations that are physically dispersed across large geographic areas; (2) varied user populations and applications; (3) multiple processor types; (4) capabilities and configurations; and (5) integration of both real-time and non-real-time operating environments within the same overall system.

The second area requires the development of mechanisms for managing all types of data both on individual hosts and across the distributed environment. To attain this capability, the critical technical challenges require: (1) developing data models and storage-and-retrieval architectures capable of handling all modalities of data in a seamless way; (2) merging and synchronizing time-dependent and non-time-dependent data; (3) developing intelligent agents capable of autonomously navigating complex database structures and extracting information for a user; (4) developing natural language and other nonparametric interfaces to support "intuitive" access and retrieval of data from the database management systems; (5) developing adaptive information distribution techniques based on context-based, as opposed to message-based, distribution; (6) using the information context for smart distribution over low-bandwidth communications in order to selectively control the quantity of information exchanged; (7) providing the capability to respond to complete information exchange failures; and (8) scaling these information distribution techniques to large systems of communications nodes.

The key to developing assured information services is adaptivity within the distributed environment. This allows a dynamic response to varying loads of crisis management or system failure, and protects the information within the system from attack or compromise. The critical technical challenges here are: (1) security mechanisms for multiclustered, real-time heterogeneous distributed environments; (2) adaptivity mechanisms that support the selective application of fault tolerance and fault avoidance strategies; (3) reconfiguration mechanisms to support graceful degradation; (4) replication mechanisms to ensure the consistency of information; (5) intelligent resource managers to dynamically respond to crisis overloads; and (6) system architectures that permit the secure use of Commercial-Off-The-Shelf (COTS) computers; software, and networks.

In the third area, development is focused on (1) real-time, heterogeneous-distributed computing environments; (2) distributed computing over high-bandwidth global grids; (3) distributed computing over low-bandwidth RF communications; (4) distributed, object-oriented, multimedia database management; (5) optimal tasking assignment to distributed resources; (6) interoperability among distributed, federated database management systems; and (7) scalability of COTS products to very large scale DoD configurations.

In the information services management area, development needs center around (1) adaptive resource management paradigms that allow dynamic reallocation of tasks to computing resources; (2) mechanisms to automatically control information exchange among nodes to limit the quantity of data based on the context of the application and available communications bandwidth; (3) mediators to assist in the acquisition of information from multiple sources within the distributed information environment; and (4) integration of both real-time and non-real-time control mechanisms within a single distributed environment.

To attain assured information services, development strategies utilize (1) extension of security mechanisms based on adaptation of commercial products to meet DoD needs; (2) development of adaptive security mechanisms that accommodate resource modifications in resource sets without violating security policy; (3) adaptive fault tolerance and avoidance mechanisms; (4) intelligent agents to dynamically respond to intermittent failures by reconfiguring the computing resource set; and (5) integrity mechanisms to ensure the validity and consistency of information in the global environment.

The DoD Advanced Logistics Program will develop and demonstrate software tools and protocols needed to gain control of the logistics pipeline. Specifically, it will produce advanced information technology to put the right materiel in the right place, at the right time, while supporting the need to do so with reduced reliance on large DoD inventories. The program plans a shared technology base of information manipulation and planning tools to support planning, execution, monitoring, and focused replanning throughout the logistics pipeline. This will be demonstrated through a system that tightly couples continuous planning and execution monitoring in an interoperable logistics support environment. The program's focus is in: (1) transportation tools to track assets and make smarter use of lift capacity; (2) rapid supply services for faster and more flexible acquisition of supplies; (3) force sustainment planning and sourcing; and (4) logistics feasibility planning that is linked to overall plans and objectives.

The need to protect proprietary information and financial data demands information system security. Here the commercial sector has capitalized on DoD investment in logistic systems and has developed commercial products for secure operating systems, secure database management systems, intrusion detection systems, and secure system design tools. Although the commercial sector has made progress enhancing the security considerations of mainstream commercial products, more remains to be done.

Several Federal and private organizations are pursuing efforts for assured information services. Both the National Security Agency and the National Institute of Standards and Technology continue to be leaders in the development of information systems security mechanisms. NASA and DOE did pioneering work in the areas of fault tolerance and high-assurance systems. A number of universities under DoD, National Science Foundation, and private sponsorship have done extensive work in fault tolerance and system integrity.

Civil Sector Enabling Research

One of the nine original NSTC committees, now restructured as a subcommittee under the Technology Committee, specifically addresses Computing, Information and Communications (CIC). Building on the Congressionally established High Performance and Computing and Communications initiative, the committee—through its CIC R&D subcommittee—coordinates a wide range of research activities. It involves 12 Federal departments and agencies, working with academia and industry. This R&D is currently being organized into five program components that are listed and described below:[11]

- High End Computing and Computation
- Large Scale Networking
- High Confidence Systems
- Human Centered Systems
- Education, Training and Human Resources

High End Computing and Computation (HECC). HECC R&D investments provide the foundation for U.S. leadership and promote the use of very high performance computing and computation for government, academia, industry, and broad societal applications. Short-term HECC development (anticipated payoffs in three to five years) addresses needs for systems software for teraflops (10^{12} floating point operations per second) by means of investments in operating systems, languages and compilers, programming environments and libraries, debugging and performance tools, scientific visualization, data management, and developments leading to a common framework and infrastructure.

Long-term HECC R&D (useful in 10 to 15 years) has helped establish scalable parallel processing as a standard for high performance computing, and has enabled the technology base for the $2 billion per year mid-range computing market. The next major long-term HECC milestone is a reliable, robust implementation of petaflops–10^{15} flops level performance and exabyte (10^{18} bytes) storage capability.

HECC investments include the infrastructure for HECC R&D, which includes the DOE national laboratories, the NASA centers, and facilities at the EPA, the National Institutes of Health, the National Oceanic and Atmospheric Administration, and the National Security Agency.

---

[11] This discussion is based upon material from the CIC Subcommittee report, *Networked Computing for the 21st Century*, August 1998.

Large Scale Networking (LSN). The LSN Program Component, including the Next Generation Internet (NGI) initiative, will help assure U.S. technological leadership in high performance network communications through research that advances the leading edge of networking technologies, services, and performance. Early Federal investments in networking R&D helped build the technological foundation of today's global Internet. Key research areas today include advanced network components and technologies for engineering and management of large scale networks of the future. LSN activities include coordinating the operation of advanced Federal networks and research addressing global-scale communications, networking security, satellite technologies, special purpose connectivity programs, and network-based applications.

The primary focus of LSN activities in FY 1999 and FY 2000 has been the Presidential NGI initiative. The NGI, in partnership with academic and industrial investments, will keep the US at the cutting edge of communications and information technologies. NGI activities are tightly coupled with the base LSN network research and infrastructure support. The NGI goals are:

- To conduct R&D in advanced end-to-end networking technologies, including differentiated services (including multicast and audio/video), network management (including allocation and sharing of bandwidth), reliability, robustness, and security.

- To prepare two prototype high performance network test beds for system scale testing of advanced technologies and services, and develop and test advanced applications. One test bed will connect at least 100 sites having end-to-end performance at least 100 times faster than today's Internet; it will be built on the Federal networks in cooperation with academic campus and regional networks. The other test bed will connect more than 10 sites having end-to-end performance at least 1,000 times faster than today's Internet.

- To develop revolutionary applications including collaboration technologies, digital libraries, distributed computing, privacy and security, and remote operation and simulation. This will have disciplinary applications in basic science, crisis management, education, the environment, Federal information services, health care, and manufacturing.

High Confidence Systems (HCS). HCS R&D focuses on the critical technologies necessary to achieve high levels of availability, reliability, restorability, protection, and security of information services. Systems that employ these technologies will be resistant to component failure and malicious manipulation and will respond to damage or perceived threat by adaptation or reconfiguration. Applications include transportation, banking, law enforcement, life- and safety-critical systems, medicine and health care, national security, power generation and distribution, and telecommunications.

FY 2000 HCS R&D includes work in assurance technologies, information security, information survivability, protecting the privacy of medical records, and secure programming languages for Internet-based applications.

Human Centered Systems (HuCS). HuCS R&D addresses increased accessibility and usability of computing systems and communications networks. Scientists, engineers, educators, students, the workforce, and the general public are all potential beneficiaries of HuCS technologies and applications.

HuCS collaboration facilities will allow researchers to conduct large-scale modeling and simulation, access appropriate information, share access and operation of remote facilities, and work within virtual environments. The results can visualize scientific data as well as configure and control experiments, regardless of geographic and temporal separation among individual members. Other HuCS R&D includes active visualization, disability and rehabilitation research, educational technologies, finding and tracking information, knowledge networks, manufacturing applications, and virtual reality.

Education, Training and Human Resources (ETHR). ETHR R&D supports computer and communications-related research to advance education and training technologies at all levels including K-12, community college, technical school, trade school, university undergraduate and graduate, and lifelong learning. The complex and technically challenging applications flowing from leading edge R&D in HECC and LSN make it increasingly important for today's students and professionals to update their education and training on an ongoing basis in order to exploit the latest technological advances. ETHR technologies improve the quality of today's science and engineering education and lead to more knowledgeable and productive citizens and Federal employees.

FY 2000 activities include new NSF centers for developing innovative learning technologies, National Institutes of Health National Medical Library and National Center for Research Resources training grants, and NASA's Web-based classroom training. Since FY 1999, NSF's Knowledge and Distributed Intelligence (KDI) initiative has focused on enabling users to access information wherever it occurs and in whatever form it is to be found. KDI will thus improve the ability to discover, collect, represent, transmit, and apply information. The result will be a greatly enhanced capability to generate, gather, model, and represent complex and cross-disciplinary scientific data and thereby create new knowledge. DoD and its Air Force Office of Scientific Research participate in ETHR through learner-centered education and automated training activities, although neither agency is part of the ETHR budget crosscut. NSF, the Department of Education, and Department of Labor (not currently a CIC agency) are addressing the need for training the workforce in information technology.

Promising Areas for Interagency and Public-Private Partnerships

Spectrum Allocation: Research in this area would provide the technical and economic knowledge needed to support policy decisions regarding allocation and efficient use of the electromagnetic spectrum and sophisticated mobile data communications technologies.

Global Positioning System (GPS) Issues and Applications: Many important transportation applications use the highly accurate GPS for position finding and navigation. One key current research topic is development of solutions to the broad financial and institutional issues associated with system growth, so that the system evolves in a manner fully reflecting the needs of civil transportation users. At the same time, there must be integration between GPS, geographic information systems (GIS) and remote sensing technologies.

Software Assurance and High-Confidence Systems: The growing complexity of intelligent systems and the greater dependence on them requires a high level of reliability, robustness, and security. Two critical areas for research partnerships are (1) system development, modeling, and verification techniques; and (2) high-confidence systems that protect and enhance the security and reliability of computer and communication networks.

ENERGY, PROPULSION, AND ENVIRONMENTAL ENGINEERING

Overview

Today's global economy has reinforced the geographic separation of production and consumption at a time when the world's population and its drive for industrial and agricultural development are increasing. The transportation challenges presented by these forces have significant environmental and energy consequences.

The economic and environmental characteristics of transportation vehicles are influenced significantly by the way that stored energy is converted into kinetic energy. A particular technology may be applicable to several modes of transportation and may be able to improve both energy efficiency and emission characteristics. Each particular alternative fuel offers strengths and weaknesses with respect to economics, practicality, and indirect impacts. Because market forces tend to promote research with near-term applications, the responsibility for exploring longer-term, higher-risk technologies and strategies to address environmental issues is met primarily through cost-shared Federal research and development.

Numerous Federal research projects currently underway seek to reduce the environmental impacts of transportation vehicles, operations, and systems. Among these projects are efforts to develop and test new energy storage and vehicle propulsion systems. Examples include fuel cells, which produce electrical energy from fuel without combustion, and flywheel batteries, which store kinetic energy directly via magnetically levitated, high-density flywheels spinning at up to 100,000 rpm. New energy storage and vehicle propulsion systems like these offer enormous potential benefits in terms of energy efficiency and emissions reductions.

Other ongoing research projects that will yield emissions and energy conservation benefits include examinations of marine vessel engine management systems, and development of hybrid bus propulsion systems (e.g., electric/diesel). Also in this category are research projects studying the emissions impacts of different motor vehicle fuels (i.e., gasoline, diesel, methanol, ethanol, compressed natural gas), and studies of the impacts of low operating temperatures on emissions of vehicles using specially formulated fuels (e.g., oxygenated gasoline).

An important ongoing project in the area of energy and propulsion involves the development of fuel cells for marine applications. A multi-agency initiative is under way to design and test fuel cells that will run on marine diesel fuel, and will provide both main vessel propulsion and shipboard auxiliary power. This technology also has important potential applications for diesel locomotives.

> *Representative Applications:*
> - **Partnership for a New Generation of Vehicles (Multiple Agencies)**
> - **Aviation Efficiency and Environmental Research (DOT/FAA, NASA)**
> - **Advanced Technology Programs (DOC)**
> - **Biofuels Feedstock Development (DOE)**
> - **Measurement and Characterization of Vehicle Particulate Emissions (EPA)**
> - **Transit Fuel Cell Propulsion (DOE, DOT/FTA)**
> - **Long-Term Highway Emissions Burden Trends (FHWA)**
> - **Hydrogen/Methanol Production Research (EPA)**
> - **High-Speed Non-Electric Locomotives (DOT/FRA)**
> - **Marine Fuel Cells (Multiple Agencies)**
> - **Advanced Vehicle Technology Programs (DOE)**

A final area of ongoing research involves improving and refining the process by which the transportation sector is monitored and regulated with respect to environmental impacts. Examples in this category include development of on-board procedures to monitor marine vessel exhaust emissions.

Illustrative Current Enabling Research

Integrated High Performance Turbine Engine Technology (IHPTET) Program.[12] The IHPTET program is an ongoing partnership led by the Defense Department with the goal of developing and demonstrating major improvements in three families of air breathing gas turbine engines that can meet future military needs. Categories being pursued include turbofan/turbojet, turboprop/turboshaft, and expendable. This program has been under way since 1988, and the partnership currently comprises six domestic turbine engine manufacturers as well as the Air Force, Navy, Army, DARPA, and NASA. The program is of substantial magnitude and is cost-shared. Annual funding is planned at approximately $150 million per year through 2003, with a comparable contribution from industry. Although the focus is military, the technologies involved are predominantly dual-use, with strong potential application to the civil sector.

The primary Defense Department objectives for the program are reduced specific fuel consumption and lowered cost for upgraded and new engines, which are also of great relevance to civil transportation. However, an additional major benefit associated with reduced fuel consumption will be proportionately lower carbon emissions. The technology base established by the IHPTET program can be expected to have widespread application to aircraft (including the high speed civil transport), ships, and electric power generation stations. The outcome should contribute significantly to reduced carbon emissions.

The Department of Energy has long recognized the potential of fuel cells for transportation applications. The DOE Fuel Cell Program's objectives are validation of fuel cell power systems that are: (a) 2-3 times more energy efficient than today's comparable vehicles; (b) more than 100 times cleaner than Federal EPA Tier II emissions standards; and (c) capable of operating on hydrogen, methanol, ethanol, natural gas, and gasoline. In addition, by 2004, the objective is to validate fuel cell propulsion systems that meet customer expectations in terms of cost

---

[12] This discussion is based on information drawn from the DOE Office of Transportation Technology Internet site (www.ott.doe.gov/oaat).

(competitive with conventional vehicles) and performance (equivalent range, safety, and reliability as conventional vehicles).

DOE is working with all stakeholders through the National Fuel Cell Alliance. This government/industry alliance includes domestic automakers, component suppliers, fuel cell developers, national laboratories, universities, and the fuels industry. Pre-competitive fuel cell R&D managed by DOE will attempt to resolve fundamental problems and issues associated with fuel cells and ancillary components that apply to a number of different fuel cell propulsion systems.

In the environmental area, NASA, DoD, and FAA—in partnership with industry and states—are researching space technologies and materials. The goal is to create innovative, environmentally friendly and higher energy propellants and longer lived power sources for space launch vehicles. These are capable of being launched any time, anywhere. For spacecraft applications, this lowers the cost of payload insertion into orbit and extends the service life for communication, navigation, and surveillance satellites.

DoD has multiple research objectives. They include: (1) improving the performance of pollution control equipment; (2) providing the capabilities to mitigate the impacts of novel new materials being adopted for advanced weapon systems; (3) minimizing environmental impacts associated with peacetime training; and when possible, (4) reducing weapon system life-cycle cost. Support facilities such as maintenance depots, shipyards, weapon stations, munitions plants, and bases require advanced pollution mitigation technologies to maintain weapon systems in mission-ready condition. Local, national, and international environmental regulations restrict military operations by increasing operating costs, reducing maintenance capabilities, and limiting training areas and opportunities.

New technologies are developed for DoD-specific problems if suitable technology is not commercially available or when the best available technologies are either too costly or fail to meet performance criteria under military operating conditions. Work is divided into four areas: cleanup, compliance, pollution prevention, and conservation. Advanced technologies in cleanup are under development to characterize and treat soils and groundwater contaminated with hazardous and toxic compounds. Contaminants of military interest are explosives, energetics, dense nonaqueous-phase liquids, and heavy metals. The objectives are to reduce cleanup costs, expedite cleanup, and ensure the protection of human health and the environment. Existing control technologies may not meet anticipated air, water, land, and noise regulations for future weapons systems. Technical efforts in compliance provide advanced "end-of-the-pipe" pollution control, treatment, recycling, and disposal technologies. Hazardous and toxic gaseous, liquid, or solid wastes are undesirable, but currently unavoidable byproducts of DoD systems, operations, and processes.

Pollution prevention complements compliance technologies. Maintenance and manufacturing processes are being developed to improve material performance and to avoid the hazardous waste and fugitive emissions generated by DoD installations, facilities, and equipment. These efforts will both reduce the burden placed on existing pollution control equipment and the overall amounts of hazardous compounds released to the environment. Soil, marine, and cultural resources at and around sea, land, and air ranges are susceptible to degradation from military operations. Work in conservation is intended to mitigate and redress impacts on DoD training ranges from readiness training and weapon development testing.

Environmental quality technical efforts are pursued collaboratively and cooperatively with EPA, DOE, NASA, USDA, academia, and private industry. The highest degree of program integration are in situ and ex situ bioremediation; DoD site characterization and analysis; DoD groundwater modeling; thermal and nonthermal plasma destruction of hazardous effluents; advanced membranes for chemical separation; specialized catalysts and regenerable chemical sorbents for air pollution control; electrochemistry; biotechnology, photolytic oxidation, sonic reaction enhancement, and supercritical water oxidation for the destruction of recalcitrant wastes; carrying-capacity models; natural resource characterization; and integrated decision support models for management of land, cultural resources, ecosystems, and threatened and endangered species.

Promising Areas for Interagency and Public-Private Partnerships

Applications of Fuel Cells: Among candidate technologies for energy storage, fuel cells offer one of the most significant potential benefits in terms of energy conversion and mitigation of adverse environmental impacts. This research would address the properties and characteristics of fuel-cell technology and potential applications to transportation vehicles.

Alternative Transportation Fuels: A variety of petroleum alternatives hold promise for transportation, each with strengths and weaknesses in terms of economics, practicality, and indirect impacts. Research partnerships in this area would explore the costs, benefits, and safety and infrastructure requirements of these various alternative fuels.

Reusable Launch Vehicles and Space Launch and Reentry Operations: Active, but fragmented partnerships exist among a number of agencies (NASA, DoD, DOC, DOT/FAA), states, universities, and commercial stakeholders towards development of low-cost launch options, high performance small satellites and hypersonic transport. The development of affordable and reliable access to space, as well as associated launch sites and operations, will enable "the high frontier" to accomplish its promise. This will result in routine space launches with impacts on communications, monitoring, navigation, and the environment.

SENSING AND MEASUREMENT

Overview

Representative Applications:
- **Highway Safety - Ice and Fog Sensors (DOT/FHWA)**
- **Smart Pavements (DOT/FHWA)**
- **Testing of Night Vision Goggles for Helicopters (DOT/USCG)**
- **Intelligent Ship Interfaces for Waterway Safety (DOT/USCG/MARAD)**
- **Track Strength Measurement Systems (DOT/FRA)**
- **Aging Aircraft Sensor and Modeling Technology (DOT/FAA, NASA)**
- **Micro-electromechanical Systems Nanotechnology Users Network (NSF)**
- **Berkeley Sensor Actuator Center, Sensing and Control of Structures (NSF)**

A wide range of information technologies are being incorporated into transportation vehicles and systems. This has increased the role of real-time monitoring and inspection of transportation vehicles and infrastructure. Numerous Federal research projects are underway or planned that focus on the development, testing, and application of sensing and measurement technologies for transportation.

Current transportation applications include development of improved roadway sensors to monitor icing, fog, and other potentially dangerous conditions. In addition, various technologies are being evaluated that will permit non-destructive inspection and monitoring of structural components such as bridge decking, beams, roadbeds, and foundations. These technologies could be coupled with communications devices along with real-time data and warning messages transmitted to maintenance and operations personnel. Proposed projects in this category include the application of sensing technologies to pavements, such that roadways would "report" damage or excessive wear to maintenance personnel.

Another category of applications addresses technologies and procedures for monitoring vehicles in real-time. For example, weigh-in-motion sensors imbedded in roadways can provide accurate readings of truck-rolling weights and assess individual vehicle compliance with weight limits, while the trucks operate at normal speed. The potential cost savings to both the public and private sector from weigh-in-motion technologies are substantial, due to the elimination of the time-consuming vehicle queues that static truck scales produce. Significant public safety advantages exist as well, since every truck can potentially be monitored, rather than just those trucks evaluated during periodic static-scale weight checks.

Improved remote sensing technologies are being developed for use in monitoring vehicle emissions in real-world field settings. These devices gather data on air pollution levels, and are essential for ongoing air quality monitoring, regulatory compliance monitoring, and assessment of pollution control strategies.

Sensing and measurement technologies are being developed and tested to improve the safety and efficiency of transportation systems. For example, airborne and satellite sensing technology under development will provide data on ocean surface currents to improve marine search and rescue operations. In addition, new surveillance capabilities under development will improve the monitoring of and response to wide-area traffic conditions, so that traffic problems can be rectified quickly.

Illustrative Current Enabling Research

DoD conducts laboratory, field, and analytical research in nondestructive structural evaluation and structural materials evaluation. Nondestructive evaluation includes: condition assessment technologies, automated signal interpretation and sensor integration, portable testing devices and systems, real-time remote-monitoring systems, and testing and calibration of sensors and systems. Structural materials evaluation includes: materials such as high-performance concrete, steel, and polymer composites for construction, repair, and rehabilitation applications, structural properties including fire endurance, standard test method development, and improved construction practices.

Promising Areas for Interagency and Public-Private Partnerships

Smart Structures and Vehicles: "Smart structures"—roads, bridges, runways, and others with a network of embedded sensors—can lower maintenance costs while improving safety and performance by continually providing detailed condition information. Likewise, "smart vehicles" sense their environmental and operating circumstances. Research in this area would identify and quantify the potential benefits and costs of coupling sensing and computing in this manner.

Micro/Nano Devices: Partnerships addressing this topic would support research and development of micro- and nano-scale devices with potential transportation applications, for example, sensors and micro-controllers for airbags, "smart" antilock brakes, engine controls, and vehicle vibration sensors.

ANALYSIS, MODELING, DESIGN, AND CONSTRUCTION TOOLS

Overview

Research in this area focuses on developing tools, knowledge, information, and techniques to improve the assessment of system requirements, design of system improvements, evaluation of alternative operational concepts and strategies, estimation of the performance likely to result from innovations, and management of system operations. Specific research areas include the following:

Transportation system design tools. These are tools and methods, such as simulations, computer models, and computer-aided design, that support systems design and process re-engineering. These tools emphasize broad system engineering and integration to assure a high level of system performance.

Research projects in this area include development of advanced simulation tools for modeling vehicle crashes, designing predictive computer models to simulate marine spills, and improvements to travel demand forecasting models.

Representative Applications:
- **Travel Model Improvement Program (DOT/EPA)**
- **Air Quality Impacts of Regional Land Use Policies (EPA)**
- **Advanced Vehicle Crash Simulation Tools (DOT/FHWA)**
- **Computer Models for Improved Spill Response (DOT/USCG)**
- **Waterway Evaluation Tool (DOT/USCG)**
- **Advanced Simulation for Monitoring and Modeling Aviation Safety (DOT/FAA, NASA)**
- **Advance Vehicles/Road Simulator (DOT/NSF)**
- **National Marine Transportation System Modeling (DOT/USCG)**
- **Transportation Analysis and Simulation System (DOT)**
- **Advisor Model for Estimating the Fuel Economy of Advanced Vehicles (DOE)**

System performance and impact characterization and modeling tools. These are means and methods by which system performance measures, such as mobility, safety, security, and economics, are assessed and integrated into system design and operational processes.

Applications in this area include the development of methods for use by states and Metropolitan Planning Organizations to monitor and forecast the effectiveness of congestion relief and mobility enhancement strategies.

Transportation and logistic system operations and management tools. Projects of this type concentrate on developing information technology and other tools to support the operation and management of transportation and logistics systems, and to assure seamless integration across various modes and organizations. Research applications in this area include the development of a computer-based Interactive Highway Safety Design Model, for considering the safety implications of highway planning and design decisions.

Transportation planning, economics, and institutions. This research involves understanding the economic, financial, and institutional context for transportation development by characterizing the needs and interests of the various transportation system stakeholders. Projects of this type include studies of the multimodal tradeoffs necessary to optimize transportation expenditures among various modes.

Illustrative Current Enabling Research

The NSF supports research addressing analytical, knowledge-based, and computational methods for modeling, simulation, optimization, and control of engineering systems. Emphasis is on development of basic methodologies, tools, and designs that are motivated by a wide variety of fundamental systems issues, including nonlinearity, scaleability, complexity, and uncertainty. The program supports leading-edge research on learning and intelligent systems, knowledge networking, neural networks, nonlinear and hybrid control, and advanced computational methods in distributed problem-solving and decision-making environments. These directions impact important industry sectors such as manufacturing and production systems, electronics, electric power, and transportation.

DoD makes extensive use of modeling and simulation tools for training, mission planning and rehearsals on synthetic battlefields, as well as for strategic planning and establishing resource requirements and acquisition programs. Core simulation technologies must provide a cost-effective and timely capability to represent systems, processes, and operational environments. Although the details are very different, the basic concepts are highly relevant to both tactical and strategic decision support for the civil transportation system. Long-term developmental objectives include the capability of full immersion of all live players into a virtual world and linkages between virtual and live instrumented simulations.[13]

DoD leads the modeling and simulation community in facilitating the interoperability of models and simulations among themselves and real-world systems. It also provides the most authoritative representations of the natural environment and systems. Many government agencies (including the Department of Transportation, Department of Justice, Federal Emergency Management Agency, and related state and local governments) participate with the Defense Department in the development of M&S standards.

DoD has the lead in the management of complex data and the development of simulations for analysis and assessment. The National Center for Atmospheric Research develops representations of weather. The private sector is addressing the modeling and simulation of individual and group behavior in terms of market research efforts and in evaluating combined human and system performance (e.g., automotive sector). DoD, other government agencies, and the commercial sector are all heavily involved in simulation interfaces (e.g., training systems, commercial entertainment interfaces, commercial design and manufacturing interfaces). The Defense Department is leveraging industry's advances in visual displays, graphics quality, and application of M&S in the design and manufacturing process.

---

[13] This discussion is based on material from the DoD Internet site at https://ca.dtic.mil/dstp.

Promising Areas for Interagency and Public-Private Partnerships

Transportation System Design Tools: Research would develop improved tools and methods that support transportation system design, with an emphasis on process re-engineering. These tools, including computer models and simulations as well as computer-aided design, would be integrated across all institutions involved in the transportation design and planning process.
Modeling and Simulation of System Performance and Impacts: Research of this type would provide the analytical methodologies and supporting data to forecast system use and impacts at the level of the individual vehicle. The resulting tools would be applicable to transportation policy development, economic planning, and impact assessment.

Transportation and Logistic System Operations and Management: In this area, research partnerships would address applications of information technology and other tools. The goal is to support the operations management of transportation systems and to assure seamless integration across organizations, modes, and institutions.

SOCIAL AND ECONOMIC POLICY ISSUES

Overview

Whether made in the public or private sector, transportation decisions must address a widening range of considerations: safety and environmental impacts, economic effects for the Nation as a whole and for various segments of the population, energy and petroleum consumption, and global competitiveness. Many stakeholders are involved, and the realities and uncertainties are complex. In particular, national policies place heavy burdens on State and local agencies for planning and decision making in many complicated areas related to transportation. There is a need for research leading to the development of a broad knowledge base that can support decision makers as they assess transportation options and alternatives.

Representative Applications:
- **Aviation Safety Risk Analysis (DOT/FAA)**
- **Capacity and Air Traffic Management (DOT/FAA)**
- **Metropolitan and Rural Policy Development (DOT/OST), DOT/FTA, DOT/FHWA)**
- **Transportation Planning and Project Development (DOT/OST, DOT/FHWA)**
- **State Planning and Research (DOT/FHWA)**

The importance of such research is clear. It has become commonplace for concerns and issues associated with social, financial, institutional, environmental and other non-technical matters to seriously compromise realization of critically needed innovations. The result is often long delays, non-optimal designs, and failure to achieve the system performance and mobility goals of the nation.

Illustrative Current Enabling Research

Policy-related research is particularly difficult to identify, since it is frequently embedded in larger technology and other programs directed toward specific objectives. The small number of activities in the RaDiUS database assigned to this category are ongoing investments within the various operating administrations of the Department of Transportation. In addition to research efforts, this topic is addressed through numerous directed programs.

Promising Areas for Interagency and Public-Private Partnerships

Transportation Planning, Economics, and Institutions: The objective of this research would be to identify and characterize the needs and interests of all parties involved with the transportation system and to understand transportation's economic, financial, and institutional context. In particular, key objectives are to find ways to increase integration of transportation modes and to identify methods to facilitate multi-modal travel.

Social Impacts of Transportation Systems and Operations: Research in this area would explore the complex relationship between transportation and society, including transportation's impact on wealth, poverty, and inequality; the social consequences of transportation innovations; transportation and race; accessibility and mobility for disabled and older persons, and the relationship between mobility and well-being.

Transportation Trends and Projections: The inherently long lifetime of transportation vehicles and infrastructure—sometimes measured in centuries—requires a forward-looking perspective in transportation planning and investment. A critical element of such projections is consideration of the tight coupling between transportation demand, and the choices available to satisfy that demand, and land-use decisions. This research activity would examine the likely consequences of current trends and project realistic future scenarios that bear on the supply of and demand for transportation services.

5. FUTURE DIRECTION AND PRIORITIES

In developing issues and ideas that relate to the future of transportation strategic research, the efforts of the Transportation Research Board (TRB) provide a valuable resource. On September 9 – 10, 1998, the TRB sponsored a workshop to examine the topic of enabling research. The participants included a balance of expertise between industry and academia as well as transportation modes; in addition, representatives from outside the transportation field were also included. The purpose was "to plan and conduct a workshop for the purpose of developing ideas and identifying opportunities for enabling research in support of the long-term goals of the nation's transportation system ..." as outlined in the *National Transportation Science and Technology Strategy*.

As part of their accomplishments, the participants examined research areas that could lead to breakthroughs and that might deserve priority funding in the future. The outcomes of the six breakout groups at that workshop are presented in the following six pages. (Social and Economic Policy Issues had not been defined as an enabling research area at that time, and so is not discussed this section.)

FOCUS AREAS FOR HUMAN PERFORMANCE AND BEHAVIOR

Topics	*Methodologies*
• **Societal Issues** – Attitudes toward alcohol – Effect of gas prices on vehicle miles – Societal attitudes toward risk	– Analysis of socioeconomic trade-offs (e.g., between safety and travel time) – Assessment of safety culture – Study of cultural and social change
• **Specific Transportation Systems** – System-induced human error – Human-centered automation and response to system failure – Flow versus accident trade-off – Technology transfer from aviation to driving	– Operator-in-the-loop modeling/simulation – Analysis of Aviation Safety Reporting system analogs (gathering of data on misses/problems) – Computer-based accident reporting – Study of human role in system reliability – Evaluation models
• **Vehicle-Operators-Environment** – Workload/task complexity and modality competition – Operator status: drowsiness, fatigue, microsleep – Alertness/attention – Visual search – Adaptive systems and operations	– Rapid prototyping – Human-computer interface and mental models – Simulation and dynamic function allocation (changing allocation of tasks to humans or machines) – Reliability analysis – Skill-, rule-, and knowledge-based models – Collection of baseline data for evaluation of new technology – Circadian rhythm/rest models
• **Operator** – Individual differences, older operators – Road rage, alcohol/drug effects – Fitness for duty/readiness to perform	– Use of advanced instructional technology – Survey/questionnaires/focus groups – Study of adaptive (to operator differences) systems

FOCUS AREAS FOR ADVANCED MATERIALS AND STRUCTURES

- **Continued research on conventional transport materials,** including cement (e.g., new cements that are less energy-intensive to produce), concrete (e.g., concrete durability, fiber-reinforced concrete), steel, aggregates, asphalt mixes. In addition, effort could be devoted to soils, an area that appears to be lacking in current research programs.

- **Materials utilization,** including adaptation of the form of structures to the materials employed, reliable joining (welding, bonding), inspection techniques, failure prediction, and recycleability of steel and pavement materials.

- **Maintenance and rehabilitation,** including use of composites in structural repair and retrofitting, cement-based materials bonding, volume stability, and fast-setting materials.

- **Provision for full- or large-scale testing of structures, pavements, and components.**

- **Research in the geotechnical area,** ground improvement/deep soil mixing, analyses and design of improved ground zones (e.g., to prevent liquefaction during earthquakes), reinforced earth structures, and site characterization (e.g., through improved geophysical techniques and the use of tomagraphy).

- **Performance-based design and specifications.**

- **Designs based on reliability/risk considerations.**

- **Continued funding of various activities already under way.** Examples are the NSF-sponsored research centers and activities associated with the Strategic Highway Research Program. Consideration could also be given to funding additional centers in specific areas, patterned after the DOE examples.

- **Expansion of research in the recycling of transport and automotive materials.**

- **Within DOT's technology transfer activities, more emphasis on the construction sector.** A possibility would be to develop a "construction extension" service.

FOCUS AREAS FOR COMPUTER, INFORMATION, AND COMMUNICATION SYSTEMS

- **Communications technologies**: wireless communications, vehicle-to-vehicle and vehicle-to-base station; driver notification systems; integration of Internet protocols into vehicles and transport systems; telecommuting and communications-assisted travel.

- **Software**: improved software requirements engineering, improved software acquisition strategies, software safety and reliability for safety-critical and high-reliability systems, improved software maintenance and management over the system life cycle.

- **Embedded systems**: on-board intelligence, vehicle systems integration, improved hardware-software synthesis, "fault-tolerant" and fail-safe systems.

- **Information systems**: driver assistance systems; data security, confidentiality, and privacy; economics of transportation-based systems; standards for and certification of interoperability; high-level requirements for transportation system management.

FOCUS AREAS FOR ENERGY, PROPULSION, AND ENVIRONMENTAL ENGINEERING

- **Energy storage** (e.g., hydrogen, batteries, flywheels).

- **Hydrogen production and distribution** (e.g., photovoltaics [solar cells converting solar energy into electricity], pipelines).

- **Hydrogen conversion on board** (reformers, fuel cells).

- **Fuel cell systems.**

- **Low CO_2 fuels** (fuels produced from water using solar energy and renewable cellulosic sources whose production and use result in reduced net carbon emissions).

- **CO_2 sequestration.**

- **Enhanced understanding and improved modeling of vehicle emissions.**

- **Ecological impacts of transportation** (protection and preservation of wetlands and other fragile ecological areas, reduced fragmentation of habitats).

- **Combustion** (catalysis, low nitrogen oxides/particulate matter).

- **Intelligent sensing/monitoring/control** (e.g., freight movements, vehicle emissions).

- **Social science** (technology acceptance/diffusion, institutional change, shifting values/culture, liability, strategies for responding to social goals).

- **Visionary concepts/designs and pathways of change** (e.g., local air delivery; avoiding collisions versus absorbing collisions; smart car sharing; land use, including livable communities; car aerodynamics).

FOCUS AREAS FOR SENSING AND MEASUREMENT

- Construction methods and equipment associated with sensing and measurement.

- Reliability, testing, modeling, and validation of sensors and sensor systems.

- Development of decision support systems relying on multiple sensing inputs.

- A systematic survey of existing sensing technologies with applications to transportation problems.

- Interagency research with NASA, EPA, and the U.S. Geological Survey to integrate satellite-based imagery and incorporate these data into an electronically accessible database. These data can be at a resolution that will support transportation planning and pre-engineering design elements to minimize environmental degradation.

- System integration procedures and policies for integrating humans into diverse sensor, communications, control, and information systems.

- A common protocol for communication among sensing and control systems.

FOCUS AREAS FOR ANALYSIS, MODELING, DESIGN, AND CONSTRUCTION TOOLS

- Standardization of frameworks for models, building on high-level architecture design developed by DOD, to link the models of different agencies.

- Efforts to ensure that models are based on relevant criteria and incorporate measures of reliability, risk, environmental impact, and social and economic metrics.

- Attempts at major modeling challenges, such as a national microsimulation model that would capture intermodal interfaces, require large-scale databases, and incorporate individual and group behavior.

- Efforts to develop models that are validated, useful, and readily accessible to users.

- Development of decision-making models to support optimization of large-scale designs and asset management.

APPENDIX A: PARTNERSHIP INITIATIVES

This Appendix contains a brief description of the Partnership Initiatives described in the *National Transportation Science and Technology Strategy*. Together with the Enabling Research presented in this document, they comprise the R&D core of the *Strategy*.

Aviation Safety Research Alliance
This initiative addresses the need to reduce the aviation accident rate as air traffic doubles over the next decade, as called for by the White House Commission on Aviation Safety and Security. Together with other partners, the FAA, NASA, and DoD will accomplish this through a coordinated program to (1) identify and conduct the research needed to meet the safety goal, and (2) work with industry to deploy research results in the form of new safety technologies.

Participants
Federal: DOD, FAA, NASA (lead agencies); DOE, NSF, NIST, NWS, and U.S. Bureau of Mines.
Other: United Nations/International Civil Aviation Organization, aircraft and avionics manufacturers, airlines, aviation organizations, universities.

Next Generation Global Air Transportation
Anticipating the future growth in air traffic, this government–industry partnership is developing the communication, navigation, and surveillance and air traffic management systems required to make "free flight" a reality. "Free flight" refers to an airspace system that greatly increases user flexibility to plan and fly preferred routes, saving both fuel and time and affording more efficient use of airspace. This activity essentially transfers the free flight concept to an operational setting prior to full deployment.

Participants
Federal: FAA and NASA (lead agencies); also DOD, NSF, NWS, USCG.
Other: Airlines, aircraft and avionics industries, academia, International Civil Aviation Organization.

Next Generation Transportation Vehicles
This partnership addresses the problems of petroleum dependence, global warming, and pollution through research leading to the development of highway vehicles, locomotives, and ships that are better designed and more efficient. It has three major thrusts: (1) continue the PNGV and Advanced Technology Transit Bus activities and supplement them by also focusing on improvements in medium- and heavy-duty-vehicle fuel efficiency; (2) support the development, test, and demonstration of non-electric high-speed rail technology; and (3) demonstrate and develop the marine application of fuel cells.

Participants
Federal: DOT (FAA, FHWA, FRA, FTA, MARAD, RSPA, USCG); DOD (Army, Navy, DARPA); DOC; DOE; and NASA—all lead agencies; also EPA and NSF.
Other: Vehicle, engine, and fuel-cell manufacturers; fuel producers; component suppliers; developers of fuel cells and other new energy-conversion technologies; shipyards; State and local authorities; universities.

National Intelligent Transportation Infrastructure

The National Intelligent Transportation Infrastructure (NITI) refers to the integrated electronics, communications, and hardware and software elements that can support intelligent transportation systems (ITS). It is a communication and information "backbone" that will enable ITS products and services to work together to save time and lives. Analogous to the local- and wide-area networks used in many workplaces, the NITI will allow surface transportation to be managed as a seamless entity by integrating transportation and management information systems across both modal and jurisdictional lines—within a region and, where appropriate, across the country.

Participants
Federal: DOT (ITS Joint Program Office—lead agency, FHWA, FRA, FTA, MARAD, USCG); DOD (USACE); DOJ (INS); Treasury (Customs); NSF; USDA.
Other: State DOTs, MPOs, emergency response and law enforcement agencies, railroads, trucking companies, information systems vendors and manufacturers, ITS Service Centers.

Intelligent Vehicle Initiative

The Intelligent Vehicle Initiative (IVI) is a government–industry program to accelerate the development and commercialization of safety- and mobility-enhancing driver-assistance systems. Overall emphasis is on four key areas: (1) evaluation of the benefits of IVI products, including collision-avoidance technologies, vision enhancements, and adaptive cruise control; (2) development of industry-wide standards for these products; (3) system prototyping; and (4) field test evaluations of the most promising products.

Participants
Federal: DOT (ITS Joint Program Office—lead agency, FHWA, FTA, NHTSA, RSPA Volpe Center); DOD (TARDEC); NSF.
Other: Motor vehicle and trucking industries, fleet operators, State and local transportation and law enforcement agencies, emergency response organizations, universities and other research organizations, professional societies.

Transportation and Sustainable Communities

This initiative explores how sustainable transportation and land use can help to achieve a balance among the often conflicting goals of economic growth, environmental quality, and sustainability. It will further Federal agencies' efforts to work with each other and with other governments, the private sector, and the public to expand understanding of the consequences of transportation choices; develop better forecasting, planning, and assessment tools; conduct technology research; and develop sustainable community and transportation initiatives.

While the primary focus of the FY 2000 Livability Initiative is to provide funding for programs directed toward mass transit, congestion relief and air quality improvement, community-based transportation programs, and "smart growth" strategies, research and development activities are supported. Therefore, the Livability Initiative will stimulate and link to a wide range of research activities.

Participants
Federal: DOD (USACE); DOE; DOT (Office of the Secretary, BTS, FAA, FHWA, FRA, FTA, RSPA); EPA; HHS (CDC); HUD; Interior (National Parks Service); OMB—all lead agencies.
Other: State and local transportation/environmental agencies and organizations; public health agencies; MPOs; mayoral offices; environmental advocates; environmental technology manufacturers and vendors; transportation system design, engineering, and construction firms; materials manufacturers; vehicle and fuel manufacturers; and universities.

Transportation Infrastructure Assurance
This partnership is developing and implementing measures to improve the security of transportation information systems, passenger and freight terminals, and other infrastructure, as well as of the people and cargo using or transiting them. It addresses (1) the physical security of transportation terminals; (2) the security of vital communication and information systems; and (3) the development and dissemination of information about security incidents and assessments of threats to transportation facilities and operations.

Participants
Federal: DOT (FAA, FHWA, FRA, FTA, ITS Joint Program Office, MARAD, RSPA, USCG); DOD; DOJ (FBI, INS, NIJ); Treasury (U.S. Customs)—all lead agencies.
Other: State and local law enforcement agencies; port and airport authorities; transportation service providers (airlines, bus lines, transit agencies, trucking companies, ship lines, railroads, parcel and freight companies).

Enhanced Goods and Freight Movement at Domestic and International Gateways
Building on earlier investments in technology, port infrastructure, and freight terminals, this partnership facilitates information exchange and technology demonstrations to promote the deployment of innovative logistics practices and information technologies at freight gateways. Initial efforts will focus on technology applications and demonstrations at the Nation's border crossings and corridors.

Participants
Federal: DOT (ITS Joint Program Office and Secretary's Office of Intermodalism—lead agencies, FAA, FHWA, FRA, MARAD, RSPA, USCG); DOC; DOD (MTMC); DOE; DOJ (INS); EPA; State; Treasury (U.S. Customs); USDA.
Other: National governments and international societies; State and local agencies; port and airport authorities; industry (air cargo companies, trucking companies, ship operators, railroads, parcel and freight companies, equipment manufacturers, and vehicle manufacturers).

Monitoring, Maintenance, and Rapid Renewal of the Physical Infrastructure
This partnership will create an environment that fosters an unprecedented level of collaboration and synergy on infrastructure research, demonstration, testing, evaluation, and technology transfer to state and local agencies. The partners will collaborate both on developing new technologies and on accelerating market acceptance of existing products.

Participants
Federal: DOT (FHWA and RSPA—lead agencies, FAA, FRA, FTA, MARAD, USCG); DOD (USACE); DOC (NIST); NSF.
Other: Civil Engineering Research Foundation; State and local agencies; chemical, automotive, and material manufacturers; commercial freight, air transport, and insurance industries; infrastructure construction, planning, and management firms; communications, water, gas, and electric utilities; universities; and industry and trade associations.

Maritime Safety Research Alliance
This partnership's focus is the prevention of maritime casualties through targeted research and development in the areas of human factors, vessel technology, and advanced information systems. It will address advanced training technologies for mariners; improved small vessel designs and structures; real-time weather systems; GPS applications; and integration of sea-based and land-based intelligent systems for traffic management and rapid emergency response.

Participants
Federal: DOT (MARAD, USCG) and DOD (MTMC, Navy, USACE)—lead agencies.
Other: Maritime industry.

Space Transportation Technology
Without affordable and reliable access to space, the future of the space program and the U.S. space transportation industry are hindered by the high cost, low reliability, and poor operability of payload launch. An unprecedented partnership between NASA and U.S. aerospace companies, this effort takes advantage of the respective strengths of government and industry by supporting NASA's efforts to develop and demonstrate pre-competitive, next-generation technology that will enable the commercial launch industry to develop full-scale, highly competitive, and reliable reusable space launchers.

Participants
Federal: FAA, NASA (lead agencies); DOD; NSF.
Other: U.S. commercial space launch providers, launch site operators, and satellite manufacturers and owners.

Accessibility for Aging and Transportation-Disadvantaged Populations
This partnership focuses on improving the mobility of the elderly and transportation-disadvantaged through better management of paratransit, advanced technologies, and livable communities. One component consists of developing, deploying, and testing a regional paratransit program that uses selected information technologies, including automatic vehicle location, geographic information systems, computer-aided dispatch, and electronic fare collection.

Participants
Federal: DOT (FTA—lead agency, Office of the Secretary, FHWA, NHTSA, ITS Joint Program Office, RSPA); DOL; EPA; HHS; HUD.
Other: State and regional agencies, MPOs, human service and employment agencies, housing authorities, public and private transit providers, information and communication system vendors, employers, universities.

Enhanced Transportation Weather Services

This partnership addresses the problems associated with adverse weather through the development of comprehensive weather information systems. One element will make use of state-of-the art weather radar, observing systems, and forecasting methods to demonstrate and evaluate an integrated weather information system—first within a "pilot" Midwestern region and eventually throughout North America. A second component is the Aviation Weather Analysis and Forecasting Program, which will improve access to and delivery of aviation weather information and reduce the consequences of weather events by generating weather observations, warnings, and forecasts with higher resolution and greater accuracy.

Participants
Federal: DOT (ITS Joint Program Office, FAA, FHWA, FRA, RSPA—lead agencies); DOD (USACE); NOAA; NWS.
Other: Iowa DOT and other State DOTs, Environment Canada, weather technology manufacturers and integrators, vehicle suppliers, ITS Service Centers, aviation industry.

APPENDIX B: LIST OF ACRONYMS

BTS	Bureau of Transportation Statistics (Department of Transportation)
CD-ROM	Compact Disk Read-Only Memory
CIC	Computing, Information and Communications
COTS	Commercial Off The Shelf
DARPA	Defense Advanced Research Projects Agency (Department of Defense)
DOC	Department of Commerce
DoD	Department of Defense
DOE	Department of Energy
DOT	Department of Transportation
EPA	Environmental Protection Agency
ETHR	Education, Training and Human Resources
FAA	Federal Aviation Administration (Department of Transportation)
FHWA	Federal Highway Administration (Department of Transportation)
FMCSA	Federal Motor Carrier Safety Administration (Department of Transportation)
FRA	Federal Railroad Administration (Department of Transportation)
FTA	Federal Transit Administration (Department of Transportation)
GIS	Geographical Information Systems
GPS	Global Positioning System
HCS	High Confidence Systems
HECC	High End Computing and Computation
HuCS	Human Centered Systems
IHPTET	Integrated High Performance Turbine Engine Technology
ITS	Intelligent Transportation Systems
ITSJPO	ITS Joint Program Office
IVI	Intelligent Vehicle Initiative
KDI	Knowledge and Distributed Intelligence initiative
LSN	Large Scale Networking
MARAD	Maritime Administration (Department of Transportation)
NASA	National Aeronautics and Space Administration
NGI	Next Generation Internet
NHTSA	National Highway Traffic Safety Administration (Department of Transportation)
NIST	National Institute of Standards and Technology (Department of Commerce)
NITI	National Intelligent Transportation Infrastructure
NOAA	National Oceanic and Atmospheric Administration (Department of Commerce)
NRC	National Research Council
NSF	National Science Foundation
NSTC	National Science and Technology Council
NTIA	National Telecommunications & Information Administration (Department of Commerce)
OMB	Office of Management and Budget

PNGV	Partnership for a New Generation of Vehicles
RaDiUS	Research and Development in the United States database
R&D	Research and Development
RSPA	Research and Special Programs Administration (Department of Transportation)
TRB	Transportation Research Board
USACE	United States Army Corps of Engineers
USCG	United States Coast Guard (Department of Transportation)
USTR	United States Trade Representative (White House)
USDA	United States Department of Agriculture

SELECTED BIBLIOGRAPHY

Papers & Reports

- National Science and Technology Council Committee on Transportation Research and Development. *Transportation and Sustainable Communities Initiative: Development of a National Investment Plan – Phase 1 Report Draft Version – 11/19/97*

- National Science and Technology Council, Committee on Transportation Research and Development, *Transportation Science and Technology Strategy*, September, 1997

- Transportation Research Board Roundtable on Research and Technology Performance Measures (November, 1996)

- Transportation Research Board Roundtable on Transportation Technologies (October, 1996).

- General Accounting Office Report: *Surface Transportation Research Funding, Federal Role, and Emerging Issues* (September, 1996)

- U.S. Congress, Office of Technology Assessment, *Defense Conversion: Redirecting R&D*, OTA-ITE-552 (Washington, DC, U.S. Government Printing Office, May, 1993).

Books

- Burris, Daniel, *Technotrends: How to Use Technology to Go Beyond Your Competition*, New York, NY: Harper Business, 1994.

- Gansler, Jacques S., *Defense Conversion: Transforming the Arsenal of Democracy*, Cambridge, MA: MIT Press, 1995.

www.ingramcontent.com/pod-product-compliance
Lightning Source LLC
Chambersburg PA
CBHW081843170526
45167CB00007B/2892